AF616794

Prentice Hall Advanced Reference Series 

Engineering

DENNO *Power System Design and Applications for Alternative Energy Sources*

FERRY, AKERS, and GREENEICH *Ultra Large Scale Integrated Microelectronics*

JOHNSON *Lectures on Adaptive Parameter Estimation*

MILUTINOVIC, ed. *Microprocessor Design for GaAs Technology*

QUACKENBUSH, BARNWELL III, and CLEMENTS *Objective Measures of Speech Quality*

Power System Design and Applications for Alternative Energy Sources

K. DENNO

Distinguished Professor
of Electrical Engineering
New Jersey Institute
of Technology

Prentice Hall, Englewood Cliffs, New Jersey 07632

Library of Congress-in-Publication Data

Denno, K. (date)
Power system design and applications for alternative energy sources.

(Prentice Hall advanced reference series)
Bibliography.
Includes index.
1. Renewable energy sources. 2. Power (Mechanics)
I. Title.
TJ808.D46 1989 621.042 88-5828
ISBN 0-13-688004-5

Editorial/production supervision
and interior design: Mary Rottino
Cover design: Diane Conner
Manufacturing buyer: Mary Ann Gloriande

Prentice Hall Advanced Reference Series

Printed in the United States of America

10 9 8 7 6 5 4 3 2 1

ISBN 0-13-688004-5

Prentice-Hall International (UK) Limited, *London*
Prentice-Hall of Australia Pty. Limited, *Sydney*
Prentice-Hall Canada Inc., *Toronto*
Prentice-Hall Hispanoamericana, S.A., *Mexico*
Prentice-Hall of India Private Limited, *New Delhi*
Prentice-Hall of Japan, Inc., *Tokyo*
Simon & Schuster Asia Pte. Ltd., *Singapore*
Editora Prentice-Hall do Brasil, Ltda., *Rio de Janeiro*

TO MY WIFE BADIA
For her love, patience, devotion,
and encouragement

Contents

Preface

This book presents for the first time the principles of system design and applications of alternative power sources that may be used in the laboratories, domestic-residential systems, and industrial grids and electric utilities generating structures.

Subject matter of this book features

1. An effective presentation of the aspects of application for alternative energy sources in the practical form of operational power systems.
2. An examination of the concept of integrating various modes of alternative energy sources in an interconnected power system, emphasizing the principles of compatibility for an effective operational power system from the standpoint of generator design, economic utilization, and ecological factors.

Chapter 1 presents electrochemical characteristics of the hydrogen and carbon fuel cells as well as their loading operations. Similar analysis is presented for several modes of storage batteries such as the lead-acid, nickel-cadmium, lithium, and sodium-sulphur batteries.

Chapter 2 presents economic optimization analysis for interconnected power system involving the integration of fuel cells, storage batteries with conventional electromechanical power sources.

Chapter 3 presents for the first time parametric as well as mathematical modeling for the process of bioelectrochemical conversion of prepared refuse into synthetic hydrocarbon fuel and then operational characteristics of the bioelectrochemical conversion of synthetic fuel to electric DC output power.

Chapter 4 presents for the first time the subject matter of the redox flow power system within the framework of its electrochemical behavior, economic feasibility, and magnetoelectric properties. Theoretical as well as experimental analysis are presented identifying classical characteristics of magnetization and relaxation of ferromagnetic fluids in the catholyte continuum.

For the important area of power electronics, Chapter 5 presents mathematical models for all types of power inverters with respect to their mode of commutation. Knowledge of inverters is essential since they are needed in transforming the DC power from all alternative energy sources into AC output for conventional application and for their integration with other AC electromechanical energy units. The remaining part of this chapter deals with the presentation of mathematical operational modes for all types of power inverters coupled to electrochemical generators. This modeling is useful in obtaining information regarding transient and steady-state performance of electrochemical sources.

Chapter 6 presents the basis of system design and application of OTEC plants (ocean thermal energy conversion) with respect to two central parameters, namely, the ammonia Reynolds number and the ocean water Reynolds number. Consideration is given to the operational procedure for the release or generation of hydrogen, oxygen, and ammonia from OTEC system.

Chapter 7 presents for the first time system design and application of the OTEC plant coupled to MHD (magnetohydrodynamic) and conventional fuel systems for the generation of electric power as well as the release of ammonia, hydrogen, and oxygen. The operational basis of endothermic as well as conventional electrolysis of water for the liberation of hydrogen and oxygen is presented. And, finally, the design principle for the interconnection of OTEC to the redox-flow-cell plant is presented showing compatibility aspects of energy storage and transport.

This textbook could be adopted by

- Undergraduate senior students in the fields of electrical engineering, mechanical engineering, chemical engineering, civil engineering, as well as engineering physics and operations research students in a technical elective course.
- Graduate students in all fields of engineering, physical sciences, and operations research, in a graduate one-semester course in the area of energy and environments.
- Consultants, practicing engineers, hobbyists, and libraries.

SPECIAL ACKNOWLEDGMENT

With special thanks and deep gratitude to Ms. Helen A. Wanner, assistant to the Dean of Engineering at New Jersey Institute of Technology, who typed and reviewed this entire manuscript with patience and dedication.

Introduction

Present modes of electric power generation are centered on the conventional electromechanical system using fossil fuel, such as oil, coal, or natural gas, and the nuclear fission system. Current research concerns the nuclear fusion program (an almost radiation-free system), which is based on the combination of unstable isotopes, mainly deterium and tritium, resulting into the release of energy after the process of fusion.

Another mode of electric power generation is the MHD-DC system using also fossil fuel such as gasified coal or preheated natural gas as the working fluid for bulk output.

In research as well as in limited applications, there are often modes of energy systems such as the conventional hydrogen or carbon fuel cell, various kinds of storage batteries, redox flow cells (oxidation-reduction of reactants), ocean thermal energy converters (solar energy extraction from the ocean, OTEC), regular solar energy converters, geothermal energy systems, as well as various modes of thermal and thermionic converter systems.

In this book system design and applications will be presented for electrochemical alternative power sources and solar energy extraction from the ocean (OTEC) as well as important characterization and modeling of electrochemical and electromagnetic phenomena involving alternative energy sources. Also, economic optimization and system coordination for the interconnection of various plants of alternative energy sources will be presented analytically, including conventional fuel cells, storage batteries, electrobiochemical cells, the redox flow cells, and OTEC plants.

Also since the regular output from those sources is direct current, detailed parametric circuit representation and modeling of solid-state inverters as well as their linkage to electrochemical generators and OTEC plants are presented.

Furthermore, because of the importance of solving the economic power coordination equation analytically, a brief presentation for the solution of partial differential equation with variable coefficients using the method of canonical system has been added as an appendix.

1

Fuel Cells and Storage Batteries

1.1 THE CONVENTIONAL FUEL CELL

1.1.1 Introduction

The principle of operation of the conventional fuel cell, which is basically the hydrogen-oxygen cell, is centered on the electrochemical conversion of oxidized fuel into electricity. Its delivered DC power output is characterized by high efficiency, effective reliability, and minimum level of emitted pollution. The fuel-cell power plant consists of three main components: the reformer, the fuel-cell stacks, and the power inverter as shown in Fig. 1.1.

The reformer is a fuel-processing component that involves the reaction of steam and mixed hydrocarbon fuel to produce hydrogen and carbon monoxide. Addition of water may follow in a second process to produce more hydrogen and carbon dioxide.

Turning to basic chemistry, we can reflect on the chemical interaction process of fuel and water in the reformer by the following:

$$C_xH_y + H_2O \rightarrow CO + H_2 \tag{1.1}$$

$$3CO + 3H_2O \rightarrow 3CO_2 + 3H_2 \tag{1.2}$$

where C_xH_y represents hydrocarbon fuel such as C_5H_{12} or C_3H_8.

The physical structure of a simple fuel cell appears in Fig. 1.2, which shows the two electrodes, namely, the fuel electrode and the air or oxidizer

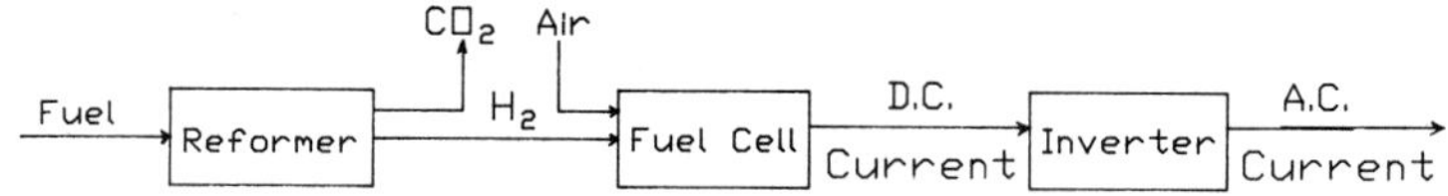

Figure 1.1 Fuel-cell unit block diagram. (Reprinted with permission of New Jersey Institute of Technology.)

electrode, with an electrolyte between them. Upon the injection of rich hydrogen at one electrode and air or oxider at the other, electrochemical reaction will commence, which produces electric power. The generating capacity of a hydrogen fuel cell is on the order of 100–200 watts at a voltage output of 1 volt. For higher output and terminal voltage, cascades of series—parallel connections of fuel cells—will meet the required electric load demands.

The third component in the fuel-cell power-processing system is the power inverter supplemented with the harmonics filter and the transformer. The function of the filter is to produce pure sinusoidal output, and the transformer will raise the output voltage to any distribution or transmission level desired. The power-processing system of the fuel-cell output is shown in Fig. 1.3.

Regarding the power inverter, which transforms the DC output from the fuel cell into an alternating sinusoidal waveform, it is structured from solid-state elements, with modes of inversions based on various operations of commutations.

Fuel-cell efficiency as established for pilot power plants is usually relatively higher than is that of the conventional electromechanical system, as shown in Fig. 1.4. Fuel-cell power-plant efficiency is essentially independent of size after an output of approximately 10 kilowatts.

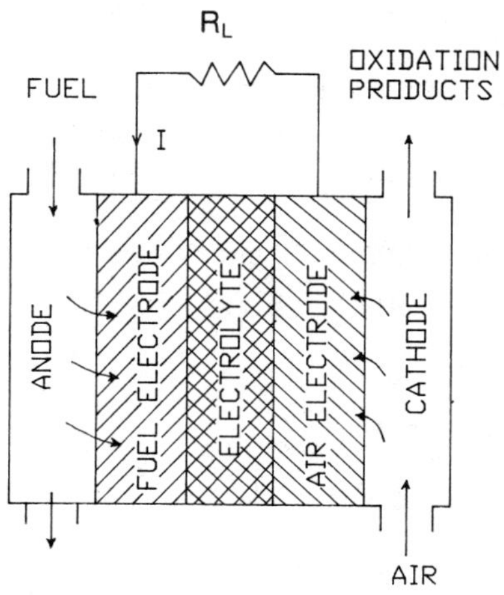

Figure 1.2 Physical elements of fuel cell. (Reprinted with permission of New Jersey Institute of Technology.)

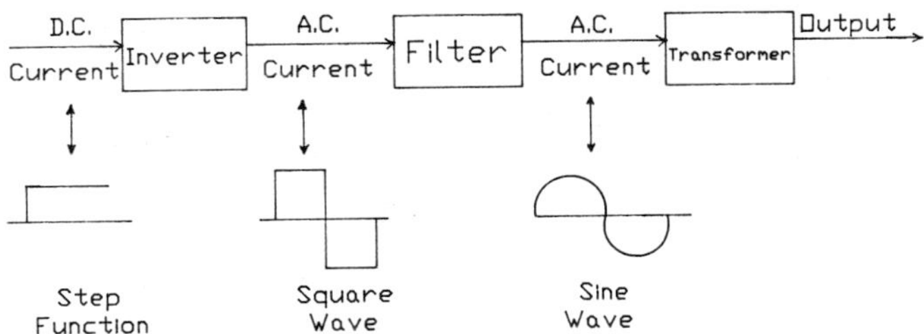

Figure 1.3 Inverter-filter-transformer block diagram. (Reprinted with permission of New Jersey Institute of Technology.)

An important correlation for the relative part-load efficiency of the fuel-cell power plant with respect to percent-rated-output power is shown in Fig. 1.5.

The fuel-cell system is completely free from human intervention after initial start-up, thus providing a major economic advantage over conventional plants in operation and maintenance cost. Because such generating units are small and almost pollution free, they can be installed near the load in a dispersed system. By doing this, the unnecessary distribution losses of energy can be minimized. Also the problems of sitting for large centralized power plant and the right of way for the additional transmission and/or distribution line construction will be minimized or in many situations be nonexistent.

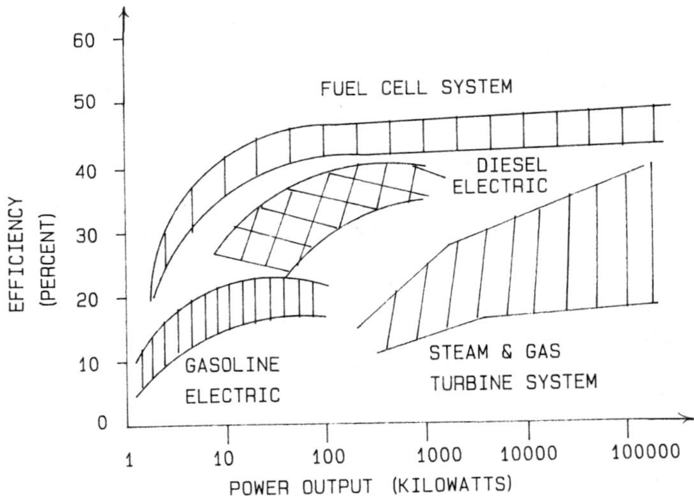

Figure 1.4 Fuel-cell efficiency characteristics. (© 1972 IEEE. Reprinted, with permission, from IEEE Transactions on Power Apparatus and Systems, Jan./Feb. 1973, pp. 230–236. Paper entitled: "Fuel-Cells for Dispersed Power Systems" by W. J. Lueckel and L. G. Eklund et al.)

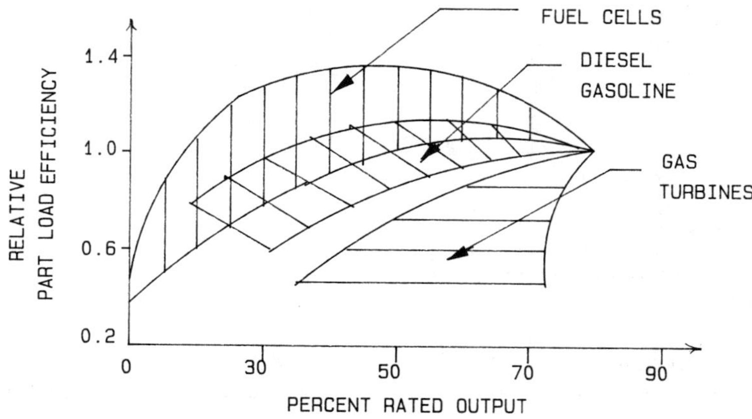

Figure 1.5 Fuel-cell efficiency versus percent-rated output. (© 1972 IEEE. Lueckel and Eklund.)

1.1.2 Electrochemical Phenomenon in the Fuel Cell

Fuel-cell performance is based on the liberation of charge carriers to the external circuit accompanied by the release of thermal energy. Since such a process depends on the stored chemical energy, the fuel cell can be considered as having very loose connections to the laws of thermodynamics.

However, the fuel-cell operation has direct reliance on Faraday's law of electrolysis, which states that 96,500 coulomb of electricity discharges 1 gram-mole per valence of an element of 96.5×10^6 coulomb/kg-mole. Hence, we can say that for an atom valency of 3, $3F$ coulomb of electricity can be generated based on Avogadro's number, which is 6.023×10^{23} electrons per kg-mole. Therefore,

$$Q = 96{,}500 \times \frac{3}{6.023} \times 10^{-23}$$

$$= 4.806 \times 10^{-19} \text{ coulomb of electricity}$$

Figure 1.6 is a simple diagram for the hydrogen fuel cell where the oxidizer is either oxygen or preheated oxygen-rich air.

At anode A,

$$2H_2 \rightarrow 4H^+ + 4e \tag{1.3}$$

At cathode K,

$$4H^- + 4e + O_2 \rightarrow 2H_2O \tag{1.4}$$

And the overall reaction is

$$2H_2 + O_2 \rightarrow 2H_2O \tag{1.5}$$

Other fuels besides H_2 are preheated air, octane, propane, and ammonia.

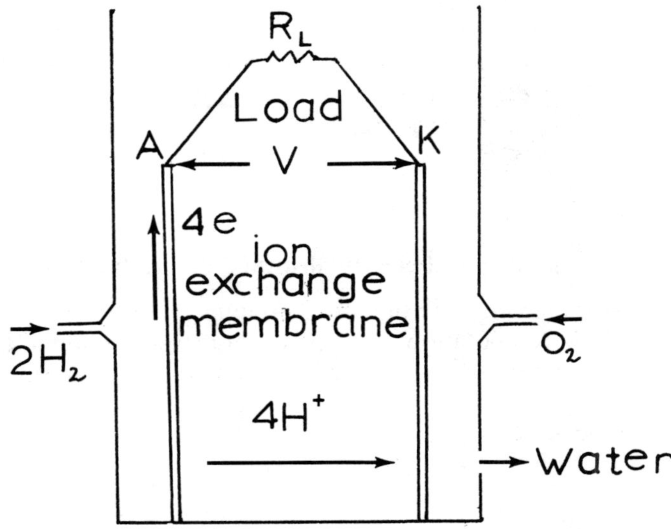

Figure 1.6 Conventional fuel-cell.

Higher-temperature cells use higher-temperature exhausts from fossil fuels instead of H_2; solid mixed oxides or fused carbonates as the selective ion membrane; and CO, CO_2, and preheated air as the oxidizer.

Throughout the chemical reaction process, the outputs will include thermal energy release, electrical energy produced, as well as mechanical energy absorbed where the reactants are gases and the product is liquid.

Therefore, (thermal + electrical) energy liberated in the fuel cell

$$= \Delta H$$

$$= \sum_{\text{reactants}} \Delta H_r - \sum_{\text{products}} \Delta H_p \tag{1.6}$$

where

H_r = enthalpy of energy formation for compound ion reactants
H_p = enthalpy of energy formation for ions products

or

$$\Delta H\Big|_{\text{anode}} + \Delta H\Big|_{\text{cathode}} = \Delta H_{\substack{\text{overall}\\\text{reaction}}} \tag{1.7}$$

Now, returning to equation 1.5, to account for heat rejection and mechanical work, we can write the modification in energy balance equation as

$$2H_2 + O_2 \rightarrow 2H_2 + 4eV + Q - 3RT \tag{1.8}$$

where

V = the cell-produced electromotive force in volts
$4eV$ = electrical energy output in joules
Q = heat energy rejected

Also,

$$3RT = P\Delta V$$

$$= \text{mechanical energy absorbed due to fluid formation}$$

where ΔV is the change in volume continuum.

For a high-temperature fuel cell where carbon monoxide acts as fuel and preheated air as the oxidizer, we can write the stages of reaction as follows:

At anode A,

$$CO + CO^{--} \rightarrow 2CO_2 + 2e \tag{1.10}$$

At cathode K,

$$CO_2 + \tfrac{1}{2}\,O_2 + 2e \rightarrow CO^{--} \tag{1.11}$$

And the overall reaction is

$$CO + \tfrac{1}{2}\,O_2 \rightarrow CO_2 \tag{1.12}$$

The enthalpy of energy formation per mole at each electrode and the sum of electrical and thermal energies released per mole in the overall reaction equation have to be specified for any fuel-cell model.

For example, perturbations in enthalpy and Gibbs energy at standard pressure of 1 atmosphere and 25°C or 298 Kelvin for carbon monoxide are -110×10^{-6} and -137.5×10^{-6}, respectively; for carbon dioxide they are -394×10^{-6} and -395×10^{-6}, respectively; for liquid water they are -286×10^{-6} and -237×10^{-6}, respectively; for steam water they are -241×10^{-6} and -228×10^{-6}, respectively; and for carbon monoxide they are -675×10^{-6} and -529×10^{-6}, respectively, where the units of the foregoing values are joule/kg-mole.[1]

1.1.3 Fuel-Cell Electric Potential

In the fuel-cell continuum, release of electrical energy is considered orderly and hence independent of change in entropy. Therefore, if n moles of electrons are associated in an electrochemical reaction producing e volts,

[1] © 1967 John Wiley & Sons, Inc. Reprinted, with permission, from the book: *DIRECT ENERGY CONVERSION,* by Edward Walsh, 1967, pp. 252.

as the internal cell voltage, the electrical energy released/kg-mole U_e is expressed by

$$U_e = nFe \tag{1.13}$$

where F is Faraday's constant.

However,

$$U_e \leq \Delta G \text{ (representing irreversibility)} \tag{1.14}$$

and ΔG is the change in Gibbs free energy. Hence from equations 1.13 and 1.14,

$$e \leq \frac{\Delta G}{nF} \tag{1.15}$$

Now we shall consider the effects of pressure and temperature on the fuel-cell potential. We can write, based on the fact of stable temperature in the fuel-cell ideal gas continuum,

$$dG = Vdp \quad \text{and} \quad \frac{PV}{T} = R \tag{1.16}$$

Therefore,

$$dG = \int_{P_1}^{P_2} RT \frac{dP}{P} = RT \ln \frac{P_2}{P_1} \tag{1.17}$$

or

$$G_2 - G_1 = \Delta G = RT \ln \frac{P_2}{P_1} \tag{1.18}$$

where

P_1, P_2 = the preceding and superseding pressure, respectively
V, T = the volume and temperature, respectively
G_1 = the free Gibbs energy at standard pressure and temperature
G_2 = the new Gibbs energy at any P and T

Hence, based on equations 1.16 and 1.17, we can write the energy balance equation represented by changes in Gibbs energy function:[2]

$$G_2 = G_1 + aRT \ln P_A + bRT \ln P_B - cRT \ln P_c - dRT \ln P_D \tag{1.19}$$

where

$a, b, c,$ and d = molar numerical concentrations
$A, B, C,$ and D = partial pressures
$P_A, P_B, P_C,$ and P_D = known as the activity of an element of reaction

[2] Ibid.

Hence, equation 1.19 could be written as

$$G_2 - G_1 = RT \ln \frac{P_A^a P_B^b}{P_C^c P_D^d} \tag{1.20}$$

And since the fuel-cell performance is toward the irreversible state,

$$U_e \leq \Delta G \tag{1.21}$$

and

$$U_e \leq nFe \tag{1.22}$$

Therefore, the fuel-cell electromotive force (emf) at any pressure and temperature is expressed by equation 1.23, which is known as the Nernst equation.[3]

$$e_2 = e_1 - \frac{RT}{nF} \ln \frac{P_C^c P_D^d}{P_A^a P_B^b} \tag{1.23}$$

We have to keep in mind that equation 1.23 is totally valid when the reactants and products obey the ideal gas laws. Therefore, its application with respect to the fuel-cell continuum, it is reasonable to say, will give approximate results, but very close to experimental verification.

Also in equation 1.23, e_1, e_2 are the fuel-cell emfs at standard pressure and temperature and new values, respectively. R is the universal gas constant and is equal to 8314 joules/kg-mole-K.

1.1.4 Fuel-Cell Ideal Efficiency

To arrive at an expression for fuel-cell efficiency, we have to proceed along a path tangential to the boundaries of the laws of thermodynamics, where under isothermal reversible conditions, the amount of useful output work is maximum and equivalent to the difference in free enthalpies between reactants and products.

Also we can state that

$$\Delta G = \Delta H - T\Delta S \tag{1.24}$$

where

ΔS = change in entropy
ΔG = maximum possible release of work
ΔH = heat of reaction
$T\Delta$ = heat exchange with the surroundings

[3] Ibid.

Therefore, from equation 1.24

$$\frac{\Delta G}{\Delta H} = \Gamma = 1 - \frac{T\Delta S}{\Delta H} \tag{1.25}$$

or Γ, the efficiency, can also be expressed by two forms:

$$\Gamma_{\text{ideal}} = \frac{nFV_t}{\Delta H} \tag{1.26}$$

$$\Gamma_{\text{actual}} = \frac{V_t it}{\Delta H} \tag{1.27}$$

where

V_t = the cell terminal voltage
i = the current drawn by the load
t = the time of discharge
n = the number of electron-mole

For the hydrogen fuel cell, values for Γ at certain absolute temperatures are[4]

T	Γ
400K	92%
500K	90%
1000K	78%
2000K	54%

Turning our attention to factors producing a reduction of the fuel cell voltage under loading conditions, we can list the following reasons:

1. Electrode polarization due to irreversibility
2. Concentration polarization
3. Ohmic polarization due to electrode and electrolyte resistances
4. Activation polarization due to the rate of oxidation

Figure 1.7 shows a plot of the fuel-cell terminal voltage versus load current with several polarization adverse effects contributing to a reduction in developed terminal voltage.

[4] © 1968 Prentice-Hall, Inc. Reprinted with permission from the book: *Direct Energy Conversion,* by S. L. Soo, 1968, pp. 54 and 56.

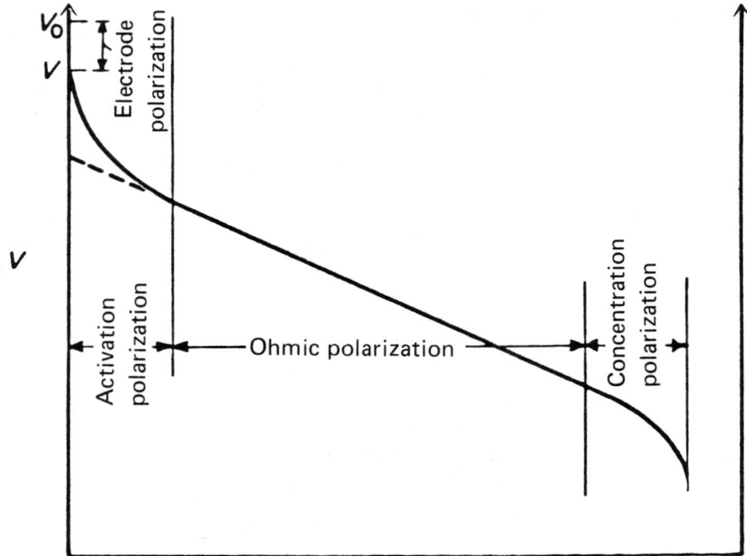

Figure 1.7 Voltage-current pattern of fuel cell. (S. L. Soo, *DIRECT ENERGY CONVERSION* © 1968 pp. 54, 56. Reprinted with permission by Prentice-Hall, Inc., Englewood Cliffs, NJ 07632)

1.2 STORAGE BATTERIES

1.2.1 Introduction

Secondary cells or storage batteries, also known as accumulators, have a high rate of electric current and are used in an environment that contains charging facilities. Well-known examples of storage batteries are the lead-acid, nickel-cadmium, and lithium cells and the sodium-sulphur batteries.

Storage batteries, usually associated with favorable initial cost, relatively a long life cycle, and reliable efficiency, in some cases are a good alternative to replace pumped storage systems in large-grid power systems. As power sources, batteries offer meaningful potential advantages when used as a separate supply station in dispersed locations, where no distribution or transmission lines are needed. As static electrochemical power sources, batteries have a relatively short lead time and a start-up time of less than 2 minutes, which is typical of pumped hydrosystems. Typical efficiency of the storage battery is on the order of 75 percent.

Utilization of the storage batteries as reliable DC power sources or for AC power output—which can be secured by its interconnection to a solid-state inverter, filter, and possibly transformer—renders attractive environmental benefits with regard to noise, undesirable emissions, and the right-of-way problems.

1.2.2 Lead-Acid Batteries

The lead-acid battery as a storage energy device was first introduced in 1860 by Gaston Plante; it is in the forefront in the main spectrum of electric energy storage systems. Its application is in starting, lighting, and ignition as well as in traction and standby power. Design improvements have led to better reliability of this battery for delivering the expected energy density under adverse weather conditions, especially at subzero temperatures where the electrochemical reactions are severely slowed down. Research and development work is progressing steadily to raise the battery nominal voltage rating energy demand and to increase life with no reduction in durability. Already the 12-volt battery has replaced the 6-volt rating with an increase of energy density from 22 to more than 40 watt hours per kilogram.

Substantial progress has been made with the maintenance-free lead-acid battery, which can be sealed and needs little or no servicing. The introduction of antimony alloyed with lead produced a remarkable improvement in the castability and physical structure of the battery cast grid. Use of antimony in the lead-acid battery structure results in sizable decline in gassing rates at normal charging voltages, with more reliable output for the open-circuit and on-charge performance periods.

Reliable and effective battery requirements are[5]

1. High-energy density
2. Efficient and rapid recharging
3. Competitive manufacturing cost
4. Long life and low maintenance
5. Long shelf life and low self-discharge
6. Good high-rate performance
7. Relatively small volume
8. Safety during an accident or charge-control failure
9. Easy replacement with little or no handling equipment

Demand for the utilization of the lead-acid battery is being extended to the area of load leveling whereby the battery could be recharged during an off-peak period of power supply. Charging of lead-acid battery could be carried out using either the constant-current or constant-voltage mode. However, regardless of the charging mode to be used, the same electro-

[5] © 1977 IEE. Reprinted, with permission, from Electronics and Power, June 1977, pp. 491–493. Paper entitled: "Developments in Lead-Acid Batteries" by M. Henderson.

chemical reactions will be involved, usually the recharge of the active mass and the conversion of water into hydrogen and oxygen.

During constant-current charging, the battery charge content and voltage increase with time as the current keeps constant level, while during constant voltage associated with current limited charging, the voltage increases under limited current control, up to the required level of voltage, where at such point the charging current declines to lower values while the voltage remains constant.

Some approximate empirical relationships reflecting on the lead-acid battery energy density and power rating are the following:

1. Energy density per cubic inch, W_1,

$$W_1 \approx 16.66t - 10.66 \tag{1.28}$$

2. Energy density per pound, W_2,

$$W_2 \approx 1.2t + 4.8, \quad \text{for } 0 \leq t \leq 6 \tag{1.29}$$

3. Power output in watts, P,

$$P \approx \frac{6}{5t + 5} \tag{1.30}$$

where t is in hours ≤ 10.

1.2.3 Nickel-Cadmium Batteries[6]

The nickel-cadmium battery is characterized as versatile, with almost no maintenance, which led to increasing reliance on its use in domestic applications as well as in security systems throughout the industry.

Versatility is secured through air oxygen recombination reaction, whereby excess gas produced by overcharge is stored inside for reuse in continuous electrochemical reaction. This is illustrated by the following chemical equation:

$$O_2 + 2H_2O + 2Cd \rightarrow 2Cd(OH)_2 \tag{1.31}$$

For the nickel-cadmium battery, discharge and charge currents are usually expressed in submultiples or multiples of 1 hour or C rate. For example, the $C/8$ will discharge the battery in 8 hours, or we can say that this battery rate is 1 amp for an 8-Ah (ampere hour) cell.

Optimum operating temperature for the nickel-cadmium battery is around 20°C, while the lowest performance temperature is −30°C and its upper limiting temperature is 60°C. Figures 1.8 through 1.12 illustrate typical operational characteristics of the nickel-cadmium cell.

[6] © 1977 IEE. Reprinted, with permission, from Electronics and Power, June 1977. Paper entitled: "Nickel-Cadmium Rechargeable Batteries" by W. D. C. Walker and C. Chem.

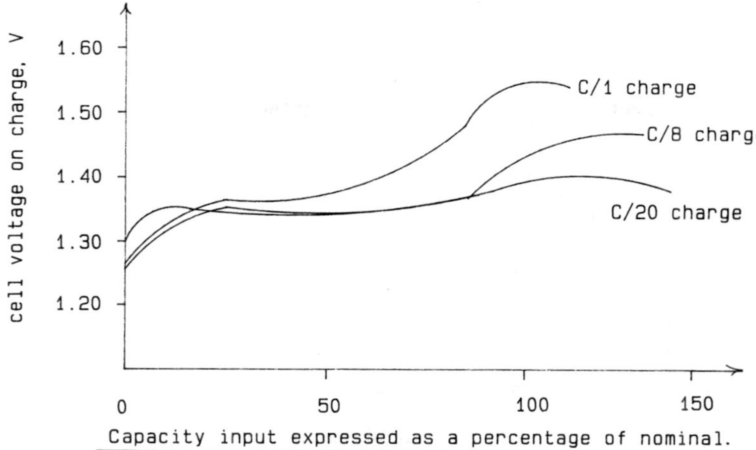

Figure 1.8 Voltage characteristic of nickel-cadmium cells under charge at an ambient temperature of 20°C. (Previously published in Electronics and Power, June 1977 by the Institution of Electrical Engineers.)

Figure 1.8 indicates that at a rate of 20 hours of discharge for a cell voltage of around 1.30 volts, the most stable performance of the nickel-cadmium cell occurs even beyond full capacity. However, when it comes to a duration of 1-hour discharge, the nickel-cadmium cell can perform at a terminal voltage exceeding 1.50 volts.

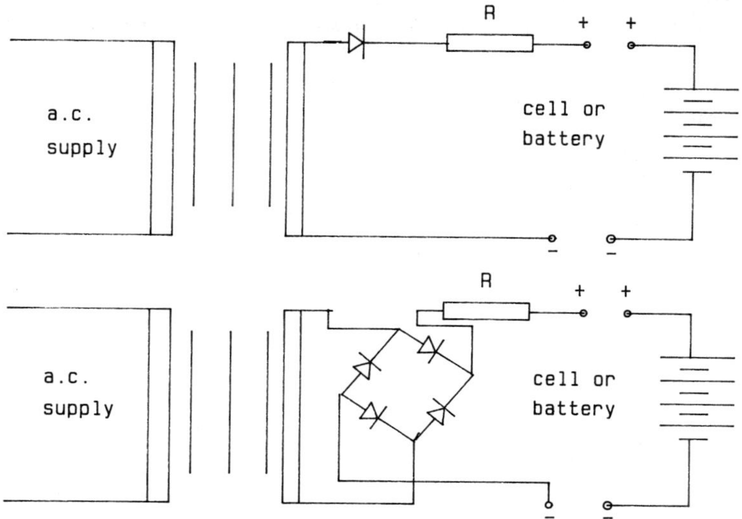

Figure 1.9 Simple charging circuits for nickel-cadmium cells. (Previously published in Electronics and Power, June 1977 by the Institution of Electrical Engineers.)

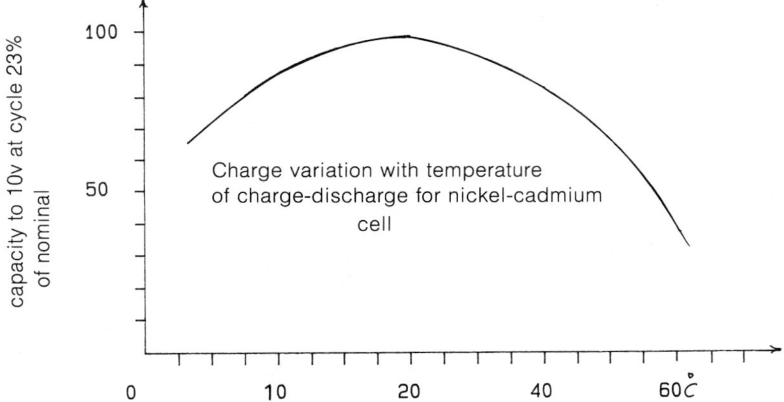

Figure 1.10 Ambient test temperature °C. (Previously published in Electronics and Power, June 1977 by the Institution of Electrical Engineers.)

Figure 1.9 presents two simple circuits for charging the nickel-cadmium cell through simple half-wave rectifier and full-wave full-rectifier circuits. Figure 1.10 shows the charge-discharge variation of the nickel-cadmium cell with respect to temperature, where at about 20°C, the cell is at its peak of voltage capacity and then declines gradually as temperature increases toward the limit of 60°C. The charging pattern of this cell climbs continuously from a low temperature of a few degrees above zero and to peak capacity at 20°C.

Figure 1.11 shows a comprehensive picture of the percentage available capacity of the nickel-cadmium cell on stand (in days) for performance at

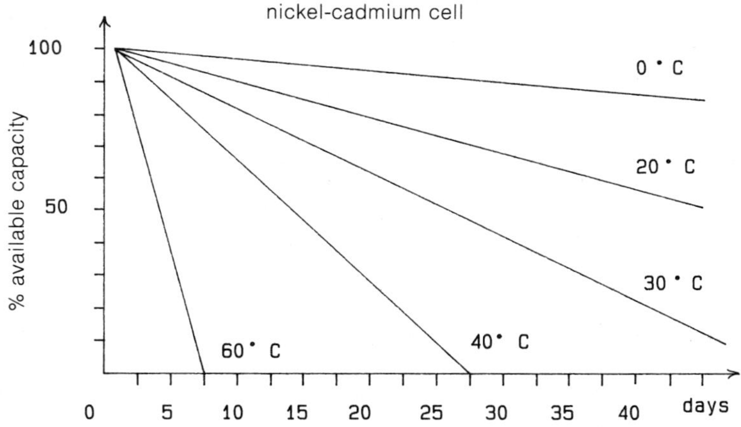

Figure 1.11 Capacity available as a function of temperature on stand after a normal charge of 12 hours at $C/8$ at 20°C. (Previously published in Electronics and Power, June 1977 by the Institution of Electrical Engineers.)

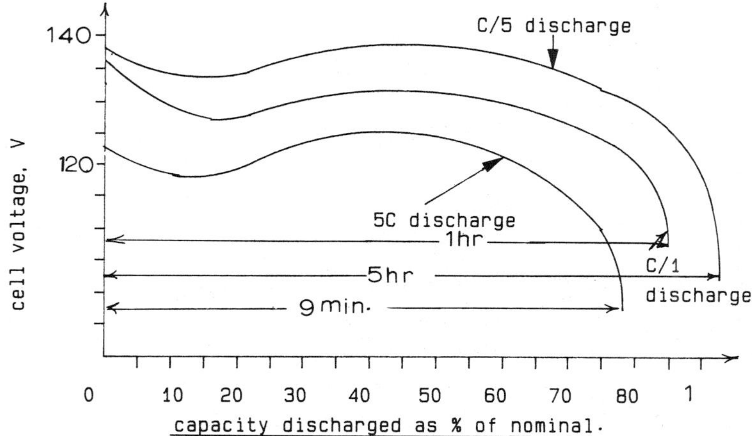

Figure 1.12 Voltage during cell discharge at 20°C for nickel-cadmium cell. (Previously published in Electronics and Power, June 1977 by the Institution of Electrical Engineers.)

several parametric temperatures ranging from 0°C to 60°C. It is remarkable that this cell can keep the most stable capacity at 0°C for a duration of several weeks, with steady declination in capacity at higher temperatures up to 60°C, where it goes down very sharply to zero in about a week.

And, finally, Fig. 1.12 illustrates patterns of voltage levels for the nickel-cadmium cell with respect to percentage capacity under three different durations of discharge, namely, 9 minutes, 1 hour, and 5 hours. The discharge was conducted at 20°C. Test conditions of the data in Fig. 1.12 are that the cells were given 23 cycles of charge-discharge with 1 discharge of $C/8$ followed by 12 hours of charge followed by a 1-hour open-circuit stand before discharge at the specified rate.

1.2.4 Lithium Batteries

Lithium is the lightest metal battery compared to other storage cells, possessing the highest electrode potential—on the order of 3.405 volt—with an ideal electrochemical equivalent of 3860 Ah/kg. Lithium–sulphur dioxide and lithium–thionyl chloride batteries can perform properly and effectively at a temperature as low as −55°C, while lithium–copper oxide batteries can operate up to 150°C temperature.

Informational data of lithium batteries with respect to their open-circuit and nominal operating voltage levels is reflected by these numbers, whereby for Li-CuO it is 240 volts and 1.50 volts; for Li-FeS_2 it is 1.8 volts and 1.5 volts; and for Li-$Bi_2PB_2O_5$ it is 1.80 volts and 1.50 volts, respectively. Other ratings include for Li-MnO_2, 3.50 volts and 3.0 volts; for

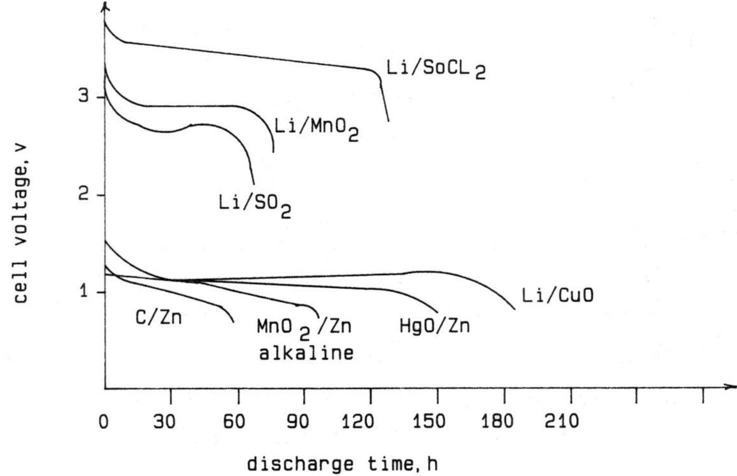

Figure 1.13 Discharge characteristics of various systems of lithium cells. (Previously published in Electronics and Power, July/August 1982 by the Institution of Electrical Engineers. © IEE.)

Li-SO_2, 2.95 volts and 2.85 volts; and for Li$SOCl_2$, 3.65 volts and 3.50 volts, respectively.[7]

We observe from the foregoing rated specifications that the Li-$SOCl_2$ battery can provide almost twice the voltage ratings of the other regular lithium batteries or any other kind of storage batteries. Figures 1.13 and 1.14 reflect the discharge characteristics of various modes of the lithium batteries.

Figure 1.13 illustrates the discharge characteristics of various modes of lithium cells in terms of their emf with respect to time in hours. We can observe that Li-$SOCl_2$ pattern of discharge shows the highest developed emf with time space of discharge close to 150 hours, while Li-MnO_2 and Li-SO_2 cells have an initial starting emf of around 3 volts, but with a total discharge time of about 75 hours. Other lithium cells in Fig. 1.13 demonstrate a stable voltage of less than 1.5 volts, but with discharge time ranging from 30 hours to 180 hours.

Turn to Fig. 1.14, which correlates voltage output under a 15-amp load with respect to days in operation for lithium–silver chromate cell, where voltage stability close to 3 volts has been maintained for about 100 days, after which a sharp decline in voltage occurred.

[7] © 1982 IEE. Reprinted, with permission, from Electronics and Power, July/August 1982. Paper entitled: "Lithium Batteries: Where the Future Lies" by Malcolm Ewing.

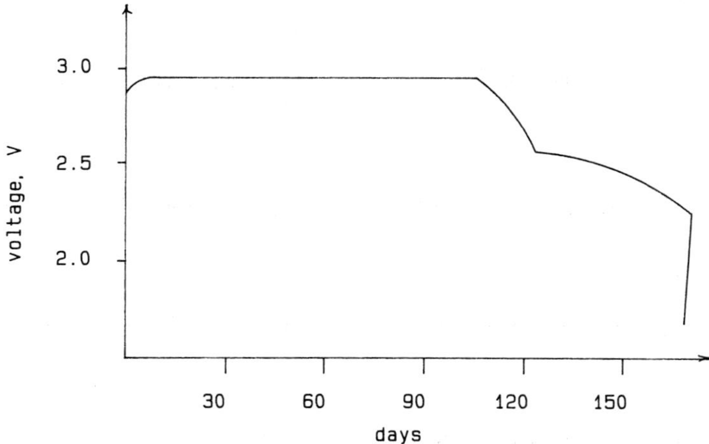

Figure 1.14 Characteristics of a lithium–silver chromate cell discharged through 15 kΩ at 37°C. (Previously published in Electronics and Power, July/August 1982 by the Institution of Electrical Engineers. © IEE.)

1.2.5 Sodium-Sulphur Batteries

Recent results in battery research and development indicate steady progress in the design of the sodium-sulphur battery with a capacity of 600 Ah for utility storage application and 150 Ah for the drive of an electric motor vehicle. However, some problems are being encountered for incorporating a large group of cells into a bulk battery assembly, so that it will possess continuous effective reliability and tolerance against thermal and vibrational perturbations.

Choride Silent Power Ltd. of the United Kingdom developed the 4-cell string series approach to develop the sodium-sulphur battery network, which incorporated a series-parallel combination of cells to guard against cells' failures as shown in Fig. 1.15. Cell failures in the sodium-sulphur cells start with a loss of resistance, followed with gradual decline in the open-circuit voltage and concludes with a rise of resistance.

The 4-cell string network developed by CSPL accommodates cell failure without resorting to the use of special protection devices. A basic principle of design for the battery network is the unit small cell, which ensures the element of continuous operation in case of failures for some cells in the network, leading to the important feature of increased reliability.

For bulk power supply, a 10-MWh sodium-sulphur battery that has been designed by CSPL is shown in Fig. 1.16. Application of the sodium-sulphur battery for powering electric motor vehicles is being given equal

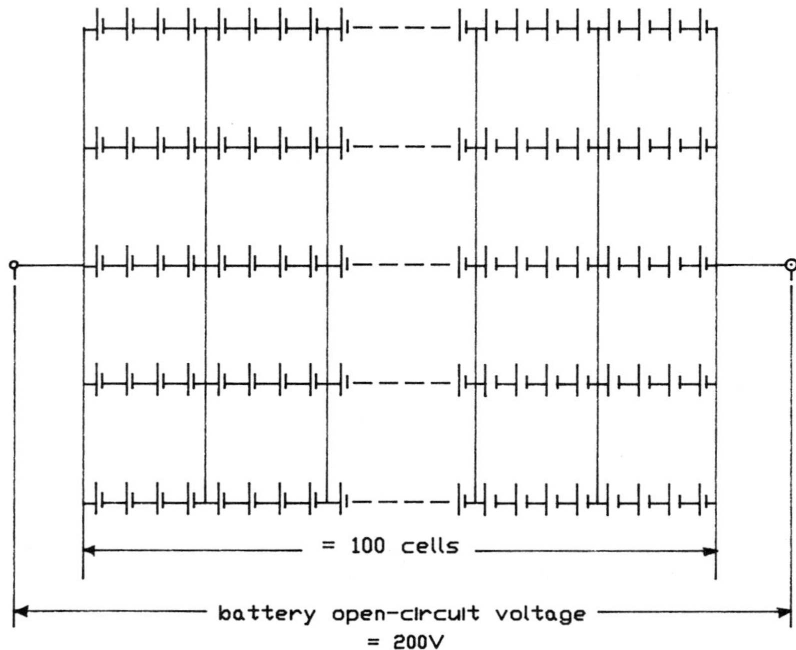

Figure 1.15 Battery network of sodium-sulphur cells. (Previously published by the Institution of Electrical Engineers. © IEE.)

Figure 1.16 Cut-away view of a beta battery load leveling installation of sodium-sulphur cells. (Previously published by the Institution of Electrical Engineers, © IEE.)

TABLE 1.1 Sodium-sulphur Vehicle Performance*

	Lead-acid	Sodium-sulphur (I)	Sodium-sulphur (II)	Sodium-sulphur (III)
Battery energy (kWh)	40	40	60	85
Range (km)	84	113	169	241
Maximum payload (t)	1.0	1.9	1.8	1.7
Battery weight (kg)	1250	330	424	580
Power available (kW, 2-h rate)	19	19	27	39

* © 1986 IEE. Reprinted, with permission, from Electronics and Power, Feb. 1986, pp. 160–162. Paper entitled: "Sodium-Sulphur Batteries, Nearing their Potential" by Peter Bindin.

priority to that for utility power supply. Differences between the established lead-acid battery and three other modes of sodium-sulphur batteries is shown in Table 1.1.[8]

1.3 SUMMARY

In this chapter basic design and operational performance principles for the fuel cell and storage battery were presented. Electrochemical reaction equations for the conventional hydrogen fuel cell and the high-temperature carbon monoxide model were discussed as was the process of fuel extraction from ordered hydrocarbon. The principle of the performance of the fuel cell was aligned through the regime of irreversibility in direct proportion with respect to the Gibbs free energy function and in reverse order with respect to the released electron modes. Expressions for the fuel-cell–produced emf in terms of reactants, products activities, and electromoles were presented as were representations for ideal and actual efficiencies.

Next, design and performance characteristics for various modes of storage batteries were presented, including the lead-acid, sodium-sulphur, nickel-cadmium, and lithium batteries. For the lead-acid battery, useful empirical rules regarding energy densities and power output with respect to the time of discharge, as well as requirements for effectiveness and reliability, were discussed. Also, operational, design, and charge-discharge characteristics were discussed for the nickel-cadmium, sodium-sulphur, and lithium batteries.

[8] © 1986 IEE. Reprinted, with permission, from Electronics and Power, Feb. 1986, pp. 160–162. Paper entitled: "Sodium-Sulphur Batteries, Nearing their Potential" by Peter Bindin.

1.4 SOLVED EXAMPLES

1. For the high-temperature carbon monoxide fuel cell, calculate the total change in enthalpy after complete electrochemical reaction. Given:

$$H_{CO} = -110 \times 10^{-6}, \quad HCO_2 = -395 \times 10^{-6}, \quad HCO^{--} = -675 \times 10^{-6}$$

all in joules/kg-mole.

Solution Refer to equations 1.7, 1.8, and 1.9 representing stages of separate reactions at the anode and cathode as well as the overall reaction.

H at the anode is

$$= H_{CO} + H_{CO--} - H_{2CO_2}$$

$$= [(-110 - 675) - (-2 \times 394)]10^6 \tag{1.a}$$

$$= 3 \times 10^6 \text{ joule/kg-mole}$$

H at the cathode is

$$= H_{CO_2} - H_{CO--}$$

$$= [-394 - (-675)]10^6 \tag{1.b}$$

$$= 281 \times 10^6 \text{ joule/kg-mole}$$

Therefore, the overall change in enthalpy is

$$H_{total} = H_{CO} - H_{CO_2}$$

$$= [-110 - (-394)]10^6 \tag{1.c}$$

$$= 284 \times 10^6 \text{ joule/kg-mole}$$

or by adding H from 1.a and 1.b, the same answer of 284×10^6 joule kg-mole is obtained.

2. Find the internal emf of H_2-O_2 fuel cell at a temperature of 50°C when preheated air is used as the oxidizer and water is the by-product. Air is at a pressure of 2 atmosphere and H_2 is at a pressure of 1 atmosphere.

Given:

$$e_o = 1.23 \text{ volts}$$

as the no-load emf.

Solution Activity of water usually is unity. For H_2 obeying the ideal gas law, its activity is represented by its pressure of 1 atmosphere partial pressure of oxygen in air $= 2 \times 0.21 = 0.42$ atm. From equation 1.19,

$$e = e_o - \frac{RT}{nF} \ln \frac{P_C^c P_D^d}{P_A^a P_B^b}$$

$e_o = 1.23$ volts. The emf at standard pressure and temperature.
From equation 1.5,

$$c \text{ or } d = 2$$

$$a = 2, \ b = 1$$

Also from equation 1.4, n for electrons = 4. Therefore,

$$e = 1.23 - \frac{8314 \times 323}{4 \times 96.5 \times 10^6} \ln \frac{1^2}{1^2 \times 0.42} = 1.136 \text{ volts}$$

3. The voltage drop of the fuel cell with chemical polarization effects e_c is expressed by the Tafel equation as

$$e_c = a + b \log_{10} J$$

where

a, b = constants that depend on electrodes material
J = the cell current density in amperes per square meter

If the internal voltage drop in the cell is linear with respect to J, and concentration polarization drop is constant, write an expression for the cell terminal voltage at any pressure and temperature and then identify the condition when $e = e_o$ and $J = 1$ KA/m^2.

Solution:

$$\begin{aligned} &\text{internal voltage drop} = e_i = c + dJ \\ &\text{chemical polarization drop} = e_c = a + b \log_{10} J \\ \therefore \quad &\text{concentration polarization} = e_n = k \end{aligned} \tag{3.a}$$

$$V_t = e - (c + dJ + a + b \log_{10} J + k) \tag{3.b}$$

$$= e_o - \frac{RT}{nF} \ln \frac{P_C^c P_D^d}{P_A^a P_B^b} - [(a + c + k) + b \log_{10} J + dJ] \tag{3.c}$$

Then for $e = e_o$, we can see from equation 3.c that for $J = 1000$ amp/m^2,

$$\frac{RT}{nF} \ln \frac{P_A^a P_B^b}{P_C^c P_D^d} = (a + c + k) + 3b = 1000d \tag{3.d}$$

4. Refer to Fig. 1.11, which represents percentage available battery capacity as a function of temperature on stand at $C/8$ at 20°C for the nickel-cadmium battery. Establish the specific set of points for the percentage capacity ν, operating temperature at the seventh day on stand.

Solution From Fig. 1.11, with a vertical line drawn from the 7.5-day point, we can read the following:

% Capacity	Temperature °C
0	60
80	40
90	30
95	20
98	0

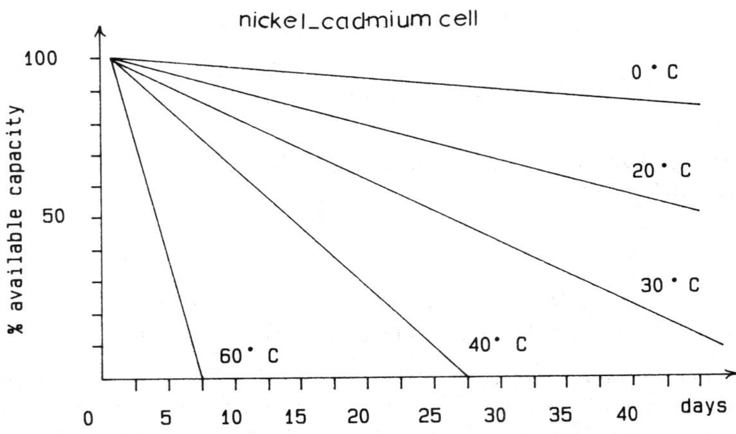

5. A lead-acid reserve cell battery has a volume of S in^3. For m units connected in series, express m and current in terms of given data for total time of h hours discharge and total power of W watts.

Given for lead-acid battery:

open-circuit voltage = 2.1 volts
watt/hr/m^3 = 2.375 watts average

Solution

$$\text{watt-hr/cell} = 2.375\ S$$

$$\text{total watts} = \frac{2.375Sm}{h} = W$$

$$\text{total terminal open-circuit voltage} = 2.1\ m\ \text{volts}$$

Therefore,

$$i = \frac{2.375Sm}{2.1mh}\ \text{Amp}$$

therefore,

$$m = \frac{Wh}{2.375\ S}\ \text{units}$$

1.5 REVIEW QUESTIONS

1. Identify the major components of reaction as well as the membrane for a high-temperature fuel cell.

2. Explain reasoning that indicates that fuel cell performance follows the laws of thermodynamics very loosely or only at the boundary.
3. Refer to equation 1.8 describing the total reaction equation for the conventional fuel cell. Identify practical problems that may arise in using such a fuel cell for space utilization and how to counter them.
4. Express the effect of reversibility and irreversibility regarding the mathematical concept of emf produced by a fuel cell.
5. Modify equation 1.23, if $P_C^c P_D^d = P_A^a P_B^b$ (P is partial pressure or activity).
6. Establish an approximate linear or piecewise linear relationship for a high-temperature fuel-cell efficiency ν, temperature.
7. Identify two methods of charging for the lead-acid battery, indicating superiority of either one.
8. Express the power output of a lead-acid battery in terms of energy density per pound.
9. Repeat Question 8 but in terms of energy density per cubic inch.
10. Compare operational performance of lithium batteries with respect to their nickel-cadmium counterpart.
11. List numerical values of electrode potential for the following modes of batteries: lead-acid, nickel-cadmium, and lithium-sulphur.
12. Identify duration (in days) of voltage stability for the lead-acid, nickel-cadmium, and the lithium batteries.
13. Identify levels of percentages-rated power output whereby efficiencies of fuel cell, diesel engine, and gas turbine coincide. Explain the superior efficiency of the fuel cell.
14. If the ideal efficiency and actual efficiency of the fuel cell are equal, express the resulting equation for the electric current density.
15. Identify possible ways to minimize the formation of water during the operation of a hydrogen fuel cell.

1.6 PROBLEMS

1. Calculate the maximum terminal voltage and the ideal efficiency of a CO-O_2 fuel cell at 600°C temperature and 30 atm. pressure.
2. For the hydrogen-oxygen fuel cell at standard pressure and temperature, calculate the total change in the enthalpy of reaction.
3. Calculate the maximum terminal voltage and the ideal efficiency of an H_2-O_2 fuel cell at 400°C temperature and 24 atm. pressure.
4. The terminal voltage of a CO-O_2 fuel cell is expressed by $V_t = (0.8 - 4 \times 10^{-4} J)$ volts. Find the current density and terminal voltage for maximum energy. Power is supplied for a duration of $2J$ seconds.
5. An H_2-O_2 fuel cell terminal voltage is expressed by the equation $V_t = (1.2 - 10^{-3} J^2)$ with effective mass density of 10 kg/m^2. If the weight density is 8 kg/m^2 of

electrode area, calculate for a total maximum power and fixed electrode area (a) V_t for each cell, (b) current density, and (c) efficiency.

6. Given the output waveform of a fuel cell coupled to a solid-state inverter as a symmetrical rectangular shape of T period in seconds, with an amplitude of A, express its expansion into the fundamental and a set of harmonics.
7. Refer to the Nernst equation for the fuel cell emf at any pressure and temperature. Establish a correlation among the reactants and product activities for $e = 0.75e_o$.
8. Consider an H_2-O_2 fuel cell where all polarization voltage reduction effects are in action. If $\Gamma_{ideal} = \Gamma_{actual}$, establish an expression for the cell current density J.
9. For a short circuit at a fuel-cell terminal, establish an expression for the current density in terms of the reactants and products activities and Tafel equation constants. Assume that only chemical polarization is present and the cell operation is in the steady state.
10. A lead-acid battery is connected across an H_2-O_2 fuel cell. If the battery terminal voltage declines linearly with respect to its current, and the fuel-cell terminal voltage declines due to chemical polarization and a constant drop due to concentration, establish a solution for the total current supplied by the combination and its efficiency at a certain load.
11. A lead-acid battery has a weight of k lb and supplies a load resistance of 1 kΩ for a time duration of 9 hours. Calculate the electric current supplied to the load.
12. For m lead-acid batteries connected in series for a duration of 1 hour, calculate the electric current supplied to a resistance load of 500 Ω and the battery assembly's actual terminal voltage at the end of 1 hour.
13. Consider a series of n lead-acid battery assemblies. Find the time at which the set will supply maximum power to a resistive load R. Consider expected discharge time of 10 hours.
14. Refer to Fig. 1.11 representing the percentage capacity of the nickel-cadmium battery V_s days on stand at several parametric temperatures. Establish correlations for the capacity V_s temperature at day 15 and then at day 25. Identify the temperature and day at which the battery capacity is maximum.
15. Refer to Figure 1.14 for the voltage discharge performance of the lithium–silver chromate battery connected across 15 kΩ resistance. Establish a new curve if the load resistance is reduced by 50 percent over the same number of days. Can you determine an approximate value for the battery internal resistance?
16. Refer to Fig. 1.13 showing discharge characteristics of various lithium batteries. Establish piecewise linear characteristics for the Li-$SOCl_2$ and Li-CuO cells. Can you establish a nonlinear empirical equation for the Li-MnO_2 cell?
17. Given Fig. 1.17 describing a curve correlating the life in years V_s number of parallel paths for the sodium-sulphur batteries, establish a possible empirical relationship representing this pattern.

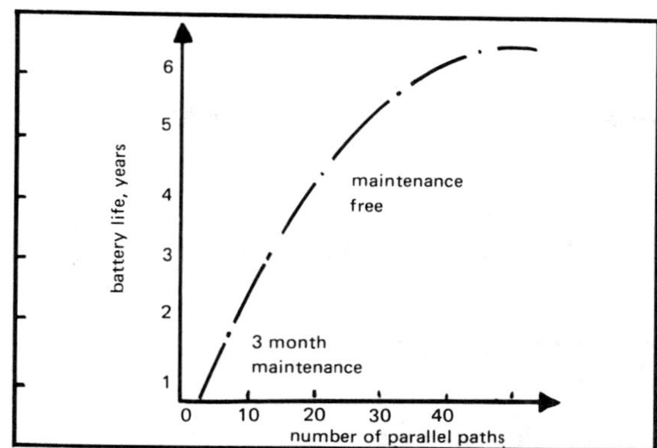

Figure 1.17 Effect of availability of parallel paths on Reliability. (Previously published by the Institution of Electrical Engineers, © IEE.)

1.7 REFERENCES

1. Bindin, Peter. "Sodium-Sulphur Batteries Nearing Their Potential." *Electronics and Power,* February 1986, pp. 160–162.
2. Crounch, Dell A., Jr., and John Werth. "Batteries and the Economics of Load Leveling in Large Power Systems." *IEEE-PAS Conference Paper* C73-213-6, January 1973.
3. Ewing, Malcolm. "Lithium Batteries—Where the Future Lies." *Electronics and Power,* July/August 1982, pp. 523–525.
4. Harrison, A. I. "Lead-Acid Standby Power Batteries in Telecommunication—Horses for Courses." *Electronics and Power,* July/August 1981, pp. 551–553.
5. Kettani, Ali. *Direct Energy Conversion.* Reading, Mass.: Addison-Wesley Publishing Company, 1970.
6. Krenz, Jerrold H. *Energy, Conversion and Utilization.* Boston: Allyn & Bacon, 1984.
7. Lueckel, W. J., and L. G. Eklund. "Fuel-Cells for Dispersed Power Systems," *IEEE-PAS* Transactions vol. 92, No. 1, January/February 1973, pp. 230–36.
8. Mantell, Charles L. *Batteries and Energy Systems.* New York: McGraw-Hill Book Co., 1970.
9. Soo, S. L. *Direct Energy Conversion.* Englewood Cliffs, N.J.: Prentice Hall, 1968.
10. Stirrup, B. N. "Charge Regimes for Nickel-Cadmium and Lead-Acid Stationary Batteries." *Electronics and Power,* July/August 1982, pp. 519–522.
11. Walker, W. D. C. "Nickel-Cadmium Rechargeable Batteries." *Electronics and Power,* June 1977, pp. 494–497.
12. Walsh, Edward M. *Energy Conversion.* New York: The Ronald Press, 1967.
13. Wheeler, Norman D. "Survey of Electrochemical Batteries." *Electro-Technology,* June 1963.

2

Modeling and Economics of Storage Batteries and Fuel Cells

2.1 MODELING OF STORAGE BATTERIES AND FUEL CELLS

2.1.1 Introduction

Due to an accelerating demand for electric energy, the continually increasing pressure of environmental forces, and the economic factors involved, power systems in the near future will be hybrids of conventional fossil fuels, nuclear plants, gas turbines, hydraulic pumped storage, storage batteries, fuel cells, and/or magnetohydrodynamic (MHD) generators. Fuel cells and storage batteries are expected to meet base loading as well as peaking power and storage power due to occasional random increases in energy demand.

Current research at various industrial, academic, and utility institutions is progressing at an increasing pace for the development of fuel cells and storage batteries to generate bulk electric power. Such power generated by electrochemical units will be transformed to AC power by solid-state power inverters of various modes of communication; then the inverter output, after adequate filtering, will be stepped up by the transformer units for the desired voltage level.

This discussion deals with the dynamic aspects of those static electrochemical generators involving chemical reaction and electric transfer of power. While the steady-state analysis of such an integrated large power system is of high importance, the aspect of dynamic behavior carries even greater emphasis to collect enough information for the behavior of the entire power system at all types of electrical, mechanical, and chemical disturbances.

To establish the required date for dynamic simulation and calculation, a complete model of the static-electrochemical generating bus must be obtained. Modeling of each of fuel cell and storage battery must center on the development of an equivalent simulation for the chemical reaction followed by electrical transform of power.

The fuel cell can be dealt with as a perfect electrochemical device on the assumptions that no variations occur in the cell, except during loading operation involving the passage of electric current, and that all changes that are associated with the flow of current can be reversed as current reverses (if reversibility is realized).

Modeling of the hydrogen-oxygen fuel cell could be considered, in general, more unusual than that of storage battery since the two basic reactants are hydrogen and oxygen. A reliable fuel cell effects the almost complete oxidation to H_2O and/or to CO_2 (for the carbon cell) of the fuel which interacts with air to generate a certain standard terminal voltage and standard current density.

The actual operating condition of the fuel cell is the irreversibility that will lead to frequent fluctuations in terminal voltage characterized by a finite difference between the maximum energy that could be developed at constant pressure and temperature and the reversible electric energy expected.

The above-mentioned deviation from reversibility will serve as a reliable criterion for the development of mathematical model for the fuel cell transient response.

In the storage battery, the basic reactants are not unique, ranging from the conventional lead-acid, nickel-cadmium, and zinc–nickel oxide to alkali-metal batteries. However, among the alkali-metal batteries, the lithium-sulphur battery (Li-S) offers promise as a reliable power source for bulk power supply, although Na-S and Li-S of the alkali-metal group have promised features close to that of (Li-S). Hence, various mathematical representations will be needed for the nature of chemical reactions corresponding to the contents of reactants.

In Fig. 2.1 a block diagram shows the location of fuel cell and storage battery on a static electrochemical bus. The intention in this section is to develop

1. Mathematical representations in the form of differential equations describing the time response of the battery, taking into account only the electrical transform of energy,
   a. Development of lumped-parameter representation of the continuum model.
   b. Development of a dynamic model for the electric transform of energy.

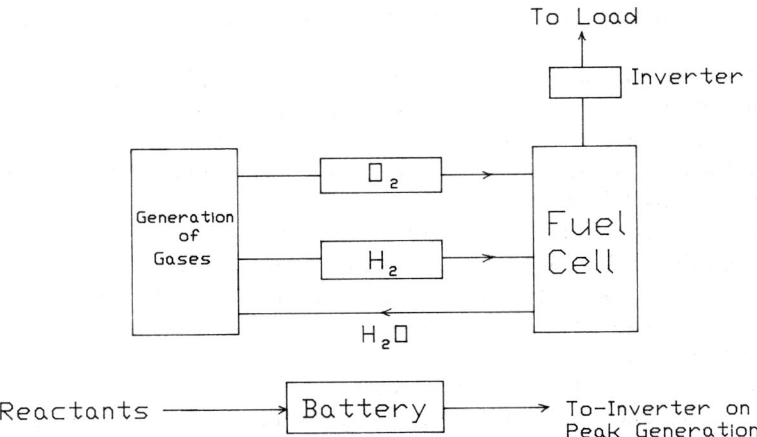

Figure 2.1 Location of fuel cell and battery in power systems. (© 1978 IEEE. Reprinted, with permission, from Proceedings—International Telephone Energy Conference #78CH1353-2, pp. 237–243, October 1978. Paper entitled: "Mathematical Modelling of Storage Battery and Fuel Cell" by K. Denno.)

2. Mathematical representation in the form of differential equations describing the time response of the fuel cell, taking into account the electrochemical transform of energy based on threshold boundary of thermodynamic principles.

  a. Development of lumped-parameter representation of the continuum model.

  b. Development of mathematical representation of the lumped-parameter model.

  c. Development of dynamic model of the electrochemical apparatus on thermodynamic basis.

2.1.2 Analogy Between Continuum and Lumped Electric Models of Storage Batteries

The general chemical structure of the DC storage battery is presented with the electrochemical equations of reactions:
The anode is denoted by A:

$$A \rightarrow A^+ + \bar{e} \tag{2.1}$$

The cathode is denoted by K:

$$K + A^+ + \bar{e} \rightarrow AK \tag{2.2}$$

The sum of equations 2.1 and 2.2 results in the overall reaction equation, which becomes

$$A + K \rightarrow AK + \text{electrical energy} \tag{2.3}$$

where AK is denoted as a reactions product, including heat release. Now, let

ρ_f = the density of charge transport in the electrolyte
$\overline{E}'$ = electric field in volt/meter or Newton/coulomb
J_f = current density due to charge transport

$$\overline{\nabla} = \overline{E}' = 0 \tag{2.4}$$

$$\overline{\nabla} \cdot \varepsilon\overline{E}' = \rho_f \tag{2.5}$$

since $\overline{D} = \varepsilon\overline{E}'$.

Then state the continuity equation

$$\overline{\nabla} \cdot \overline{J}_f + \frac{\partial \rho_f}{\partial t} = 0 \tag{2.6}$$

where

$$\overline{J}_f = \sigma\overline{E}' \text{ (for stationary material motion)} \tag{2.7}$$

Therefore, equation 2.6 becomes

$$\sigma\overline{E}' + \frac{\partial}{\partial t}[\overline{\nabla} \cdot \varepsilon\overline{E}'] = 0 \tag{2.8}$$

Next, let

$$\overline{E}' = -\overline{\nabla}E \tag{2.9}$$

where E = scalar potential function across the battery continuum. Therefore,

$$\overline{\nabla} \cdot \sigma\overline{\nabla}E = -\frac{\partial}{\partial t}\overline{\nabla} \cdot \varepsilon\overline{\nabla}E \tag{2.10}$$

From equations 2.6 and 2.7

$$\frac{\sigma}{\sigma}\rho_f + \frac{\partial \rho_f}{\partial t} = 0 \tag{2.11}$$

where the general solution of equation 2.11 is in the form

$$\rho_f(x, y, z, t) = \rho_o(x, y, z, t)\overline{e}^{t/\tau} \tag{2.12}$$

τ is known as the relaxation time constant, expressed by

$$\tau = \frac{\varepsilon}{\sigma} \tag{2.13}$$

Also, from equation 2.8, we can conclude that

$$\overline{\nabla} \cdot \sigma\overline{E}' + \overline{\nabla} \cdot \frac{\partial}{\partial t}\varepsilon\overline{E}' = 0 \tag{2.14}$$

Therefore,

$$\sigma\overline{E}' = -\frac{\partial}{\partial t}\varepsilon\overline{E}' \tag{2.15}$$

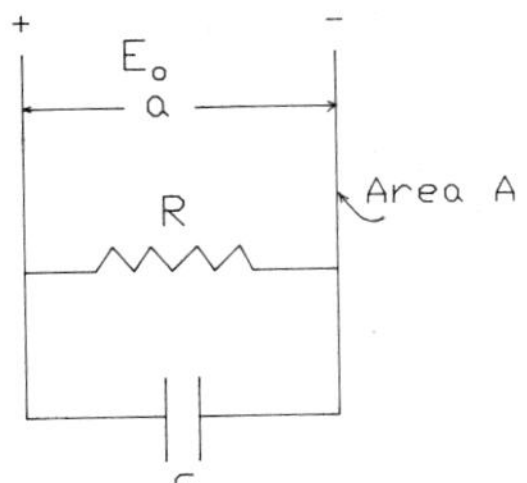

Figure 2.2 Circuit model of storage battery. (© 1978 IEEE. Reprinted, with permission, from INTELEC '78.)

A solution to equation 2.15 with initial $E'_o = E_o/a$ can be written as

$$E' = \frac{E_o}{a} \bar{e}^{t/\tau} \tag{2.16}$$

which is the transient behavior of the electric field.

Solutions presented in equations 2.12 and 2.16 resemble the behavior of charge transport and field decay in a continuum model of the battery system. Now we turn our attention to the storage battery.

A system shown in Fig. 2.2 comprising a capacitance with its dielectric loss, C and R, respectively, have a transient behavior similar to that of the continuum model just established. With

$$\tau = RC, \tag{2.17}$$

we can see from the physical and geometrical representation of the capacitance that

$$C = E\frac{A}{a}$$

$$R = \frac{1}{\sigma}\frac{a}{A}$$

Therefore,

$$RC = \frac{A}{a}\frac{1}{\sigma}\frac{a}{A} = \frac{\varepsilon}{\sigma} \tag{2.18}$$

2.1.3 General Mathematical Model of the Storage Battery

Based on lumped-circuit representation and from the fact that a capacitor with initial energy storage could be represented as in Fig. 2.3 in the S domain as either a or b, where

$S = \alpha + jW$, known as the complex frequency
$\alpha =$ the rate of attenuation
$W =$ the angular frequency

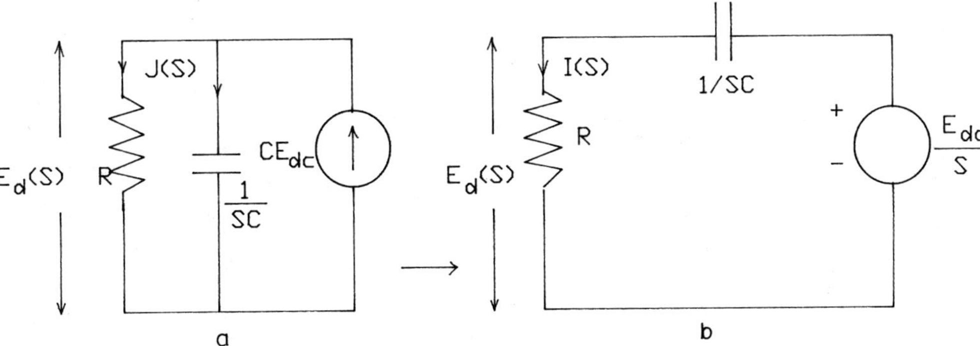

Figure 2.3 Representation of charged capacitor in the frequency domain. (© 1978 IEEE. Reprinted with permission, from INTELEC '78.)

The lumped equivalent circuit of a storage battery in the complex frequency domain can be in the form already shown in Fig. 2.3, where

$$E_{do}(S) = \frac{E_{do}}{S} \tag{2.19}$$

and

E_{do} = the initial DC terminal voltage at battery terminals

Therefore,

$$I(S) = \frac{E_{do/S}}{R + 1/Sc} \tag{2.20}$$

and

$$\begin{aligned} E_d(S) &= \frac{R}{R + 1/Sc} \frac{E_{do}}{S} \\ &= E_{do} \frac{RC}{1 + RCS} \end{aligned} \tag{2.21}$$

And we can express

$$\begin{aligned} \frac{E_d(S)}{E_{do}} &= \frac{RC}{1 + RCS} \\ &= \frac{G}{1 + GH} \end{aligned} \tag{2.22}$$

where

$$G = RC$$

$$H = S$$

Therefore, the mathematical model based on charge transport phenomenon can be structured as shown in Fig. 2.4. Now since $RC = \tau$,

$$\frac{E_d(S)}{E_{do}} = \frac{\tau}{1 + \tau S} \tag{2.23}$$

However, it remains to develop a transfer function representing the equivalent chemical reaction for particular types of battery reactants.

2.1.4 Lumped-Parameter Electric Model of the Fuel Cell

As a review from chapter 1, we can restate that the operation of the fuel is characterized by the general equation of reaction:

$$\text{fuel} + \text{oxygen (as oxidizer)} \rightarrow \text{products} \tag{2.24}$$

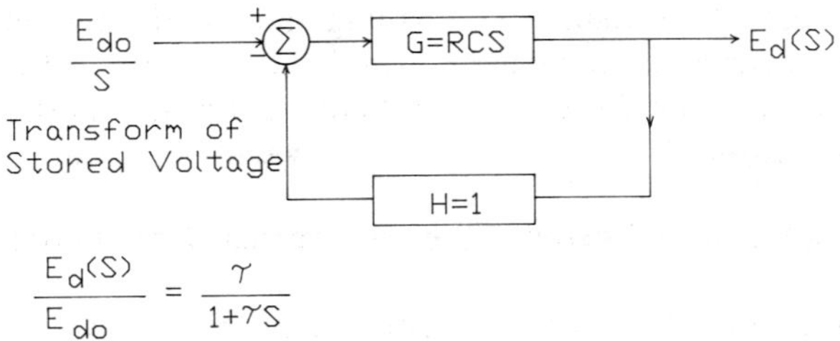

Figure 2.4 Mathematical model of battery based on electric charge transport. (© 1978 IEEE. Reprinted with permission, from INTELEC '78.)

Let us look at Fig. 2.5, which illustrates the simple charge transport phenomenon that is expressed by the following equations of reaction:

The anode reaction is

$$2H_2 \rightarrow 4H^+ + 4\bar{e} \tag{2.25}$$

The cathode reaction is

$$O_2 + 4H^+ + 4\bar{e} \rightarrow H_2O \tag{2.26}$$

or

$$2H_2 + O_2 \rightarrow 2H_2O + 4eE + Q - 3RT \tag{2.27}$$

where

Q = molar heat energy released
$4eE$ = electric energy released
$3RT$ = molar mechanical energy absorbed in the process
R = the universal gas constant = 8314 joule/kg-mole-K
E = generated cell voltage or emf

We see in Fig. 2.5 that hydrogen ions diffuse through the ion exchange membrane separating the two electrodes. Before connecting the load across anode and cathode, an initial charge of n_e is cumulative at the cathode with

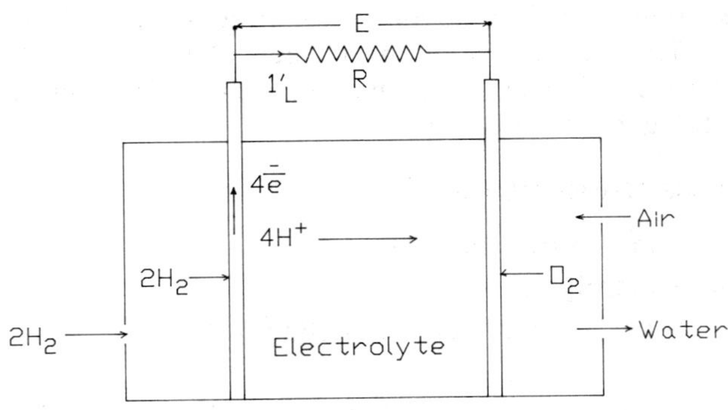

Figure 2.5 Physical feature of a fuel cell. (© 1978 IEEE. Reprinted, with permission, from INTELEC '78.)

the establishment of an open-circuit voltage across the two electrodes. On this basis, an electrical model can be established exactly as that deduced for the storage battery; and hence the mathematical model is the same as that identified in equation 2.18.

2.1.5 Fuel Cell as Perfect Electrochemical Apparatus

As we mentioned earlier, performance of the fuel cell is based on the completion of the chemical reaction of fuel to be oxidized with oxygen or preheated air (for a high-temperature cell).

A reasonably efficient fuel cell involves the oxidization of fuel to H_2O or CO_2 since the interaction with air or oxygen will deliver basic terminal voltage with reasonable current density on the order of 100 ma/cm^2 corresponding to a terminal voltage of the order of 0.60 V. Energy loss in the form of I^2R can be minimized by reducing electrode spacing and with operational electrolyte having high ionic level and very small electronic conductivity.

Perfection of the electrochemical reaction is based on the assumption that no internal variations take place in the cell, except that all changes that are associated with the flow of current can be reversed by reversing that current.

There are two basic requirements for a perfect fuel cell: these are reactivity and invariance.

REACTIVITY REQUIREMENTS[1]

1. *Stoichiometric condition*: To obtain the maximum amount of electricity from oxidization of 1 mole of fuel, H_2O or CO_2 must be the product of reaction since they represent the oxidization product of lowest energy. For example, oxidation of [C] to [CO] will generate 2 Faradays of electricity per mole of fuel instead of 4 Faradays if oxidized to CO_2.
2. *Kinetic condition*: This implies smooth coupling with high-electrode activity. The rate of liberation of valence electrons to load depends upon the rate of reaction at the electrodes. Consequently, this will establish higher terminal voltage.

INVARIANCE REQUIREMENTS[2]

1. No corrosion or side reactions.
2. Invariant electrolyte.
3. No serious variation in electrodes composition.

[1] H. A. Liebhafsky and E. J. Cairns, *Fuel Cells and Fuel Batteries* (New York, N.Y.: John Wiley & Sons, Inc. 1968).

[2] Ibid.

However, a chemical reaction in the fuel cell will generate products other than CO_2 and H_2O, leading to the conclusion that N *will not be an integer*, where

N = Equivalents of chemical changes that occur during oxidization process

Also, practical operation of the fuel cell is irreversible, rendering a sizable voltage change due to actual difference between Gibbs free energy and the reversible electrical energy as expressed by the relationship

$$\frac{|\Delta G| - |W_e|}{nF} = \Delta E \tag{2.28}$$

where

ΔG = Gibbs free energy, defined as maximum energy available at constant P and T (pressure and temperature)
$\Delta G \geq neF$ for electrical energy release
W_e = reversible electric energy = NFE
n = molar concentration
F = the amount of electricity represented by one chemical equivalent = 96,493 coulombs
N = equivalent orders of electrochemical reactions
E = change in cell terminal voltage from no-load value

Now, we can write

$$\Delta G = \Delta H - T\Delta S \tag{2.29}$$

where

ΔH = work exchanged with atmosphere
$T\Delta S$ = work exchanged with the environment

Of course, we are assuming that our analysis of the fuel cell is occurring at the boundary of the law of thermodynamics.

2.1.6 Dynamic Model of the Fuel Cell on Thermochemical Basis[3]

Let, ΔG be defined as the maximum energy available from a chemical reaction at constant pressure and temperature. It is expressed as a sum of individual chemical potentials of all reactants:

$$\Delta G = \sum_{i=1}^{i=j} \mu_i dn_i \tag{2.30}$$

[3] K. Denno, "Steady-State and Dynamic Investigations for Determining Optimum Electrochemical-Electromechanical Interconnected Power System," Technical Report of June 1974, Energy Utilization of PSE&G, NJ.

where μ_i is the electrochemical potential of the ith reactant. The overall energy equation can be expressed by

$$dU = TdQ - dG \tag{2.31}$$

where dU is the total energy in joules.

However, irreversibility is an actual feature of the fuel-cell operation where

$$nFE = G - NFE \tag{2.32}$$

or differentiating,

$$dG = F[ndE - Edn] - F[NdE + EdN] \tag{2.33}$$

From equations 2.32 and 2.33, inserting time variation and without volume change will result in

$$\frac{du}{dt} = F\left(n\frac{dE}{dt} + E\frac{dn}{dt}\right) + F\left(N\frac{dF}{dt} + E\frac{dN}{dt}\right) + TdQ \tag{2.34}$$

where s is the entropy of reactants.

Taking the Laplace transform of equation 2.34, as shown in Appendix B, will result in the following equation in the complex frequency domain:

$$\frac{E(S)}{U(S)} = G - \frac{U_o/S}{U(S)}G + \frac{E_o/S}{U(S)}G' + \frac{1/FS}{U(S)}G'' \tag{2.35}$$

The mathematical model based on equation 2.25 is shown in Fig. 2.6, where

$N(S)$ = transform of a function representing the equivalency of an ordered chemical reaction

$n(S)$ = transform of a molar concentration function for the reactants

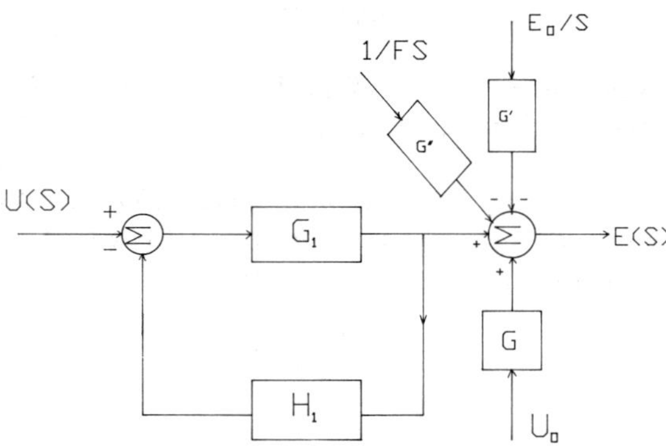

Figure 2.6 Mathematical model of fuel cell on thermodynamic principles. (Reprinted with permission of Energy Utilization of P.S.E.&G., NJ.)

2.1.7 Appendix[4]

The electrochemical differential equation describing the operation of the fuel cell is

$$\frac{dU}{dt} = T\frac{dQ}{dt} + F\left(n\frac{dE}{dt} + E\frac{dn}{dt}\right) + F\left(N\frac{dE}{dt} + E\frac{dn}{dt}\right) \qquad (2.36)$$

where Q is the heat of reaction released.

Using the following relationship for Laplace transformation of the product of two functions,

$$[f_1(t)\, f_2(t)] = \sum_{k=1}^{g} \frac{A_1(S_k)}{B_1'(S_k)} F_2(S - S_k) \qquad (2.37)$$

where

$$F_1(S) = \frac{A_1(S_k)}{B_1(S)}$$

and is assumed to have first-order pole

$$S_k = \text{the } k\text{th-order pole of } F_1(S)$$

Now let

$$n(S) = \frac{A_1(S_{k_1})}{B_1(S_{k_1})}$$

$$N(S) = \frac{A_2(S_{k_2})}{B_2(S_{k_2})}$$

$$Q(S) = \frac{A_3(S_{k_3})}{B_3(S_{k_3})} \qquad (2.38)$$

and

$$E(S) = \frac{E}{S}$$

Then using equation 2.36, the following transform equation is obtained with

$$E(S) = G - \frac{U_o/S}{U(S)}G + \frac{E_o/S}{U(S)}G' - \frac{1/FS}{U(S)}G'' \qquad (2.39)$$

$$G = \frac{1/S[n(S) = N(S)]}{\dfrac{1 + \sum\limits_{k_1}^{q_1} [A_1(S_{k_1})/B_1'(S_{k_1})] + \sum\limits_{k_2}^{q_1} [A_2(S_{k_2})/B_2(S_{k_2})]}{S[n(S) = N(S)]}} \qquad (2.40)$$

$$G = \frac{G_1}{1 + G_1H_1}$$

$$G_1 = \frac{1}{F}\frac{1}{S[n(S) = N(S)]} \qquad (2.41)$$

[4] Ibid.

$$H_1 = F\left[\sum_{k_1}^{q_1} \frac{A_1(S_{k_1})}{B_1'(S_{k_1})} + \sum_{k_2}^{q_2} \frac{A_2(S_{k_2})}{B_2'(S_{k_2})}\right] \tag{2.42}$$

$$G' = \frac{\dfrac{\sum_{k_1}^{q_1} [A_1(S_{k_1})/B_1'(S_{k_1})] + \sum_{k_2}^{q_2} [A_2(S_{k_2})/B_2'(S_{k_2})]}{S[n(S) + N(S)]}}{1 + \dfrac{\sum_{k_1} A_1(S_{k_1})/B_i(S_{k_i}) + \sum_{k_2} A_2(S_{k_2})/B_i(S_{k_2})]}{S[n(S) = N(S)]}} \tag{2.43}$$

$$= \frac{G_i'}{1 + G_1'H_1'}$$

$$G_1' = \frac{\sum_{k_1}^{q_1} [A_1(S_{k_1})/B_1'(S_{k_1})] + \sum_{k_2}^{q_2} [A_2(S_{k_2})/B_2'(S_{k_2})]}{S[n(S) + N(S)]} \tag{2.44}$$

$$H_1' = 1$$

$$G'' = \frac{\dfrac{\sum_{k_3}^{q_3} [A_3(S_{k_3})/B_3'(S_{k_3})]\, T(S - S_{k_3})}{\sum_{k_1}^{q_1} [A_1(S_{k_1})/B_i(S_{k_1})] + \sum_{k_2}^{q_1} [A_2(S_{k_1})/B_2'(S_{k_2})]}}{1 + \dfrac{S[n(S) + N(S)]}{\sum_{k_1}^{q_1} [A_1(S_{k_1})/B_1'(S_{k_1})] + \sum_{k_2}^{q_2} [A_2(S_{k_2})/B_2'(S_{k_2})]}} \tag{2.45}$$

with

$$G_1'' = \frac{\sum_{k_3}^{q_3} [A_3(S_{k_3})/B_3'(S_{k_3})]\, T(S - S_{k_3})}{\sum_{k_1}^{q_1} [A_1(S_{k_1})/B_2'(S_{k_1})] + \sum_{k_2}^{q_2} [A_1(S_{k_2})/B_2'(S_{k_2})]} \tag{2.46}$$

$$H_1'' = S[n(S) + N(S)] \sum_{k_3}^{q_1} \frac{A_3(S_{k_3})}{B_3'(S_{k_3})}\, T(S - S_{k_3}) \tag{2.47}$$

2.2 ECONOMIC FEASIBILITY OF STORAGE BATTERIES AND FUEL CELLS IN A GRID POWER SYSTEM

2.2.1 Introduction[5]

In this discussion, storage batteries and conventional (hydrogen) fuel cells will be integrated in an interconnected grid power network containing

[5] Ibid.

TABLE 2.1 Operation and Maintenance Costs for Various Types of Generating Units*

Unit Type	Average Life (years)	Fixed Operation and Maintenance ($/kw/year)	Variable Operation and Maintenance (mills/kWh)	Annual Maintenance (weeks)
Steam fossil	50	4.0	0.6	4
Gas turbine	30	negligible	4.0	2
Nuclear	30	2.0	0.2	5
Pumped hydro	50	negligible	negligible	5
Fuel cell	20	negligible	3.0	1
Storage battery	20	negligible	3.0	1

(* Reprinted with permission of Energy Utilization of P.S.E.&G., NJ.)

power-plant sources of fossil steam, gas turbine, nuclear, and pumped hydroelectrics. The analysis is geared to present the economic feasibility of storage batteries and fuel cells in meeting the demand of peak loading as well as base loads in the new integrated power system network.

Figures reflecting on the initial cost of storage batteries are approximately around $400/kW for unit capacity of less than 0.5 MW, $350/kW for unit capacity of 0.5–2 MW, and $300/kW for unit capacity of 2–500 MW. For the fuel cell, initial costs are approximately $400/kW for unit capacity of less than 0.5 MW, $350/kW for unit capacity of 0.5–2 MW, and $300/kW for unit capacity of 2–5 MW.

In addition to those guiding figures for the storage batteries and fuel cells, the data listed in Table 2.1 represent operation and maintenance cost for various types of generating units.

2.2.2 The Process of Economic Allocation of Power[6]

This process is centered on solving the economic coordination equation:

$$\frac{dF_n}{dP_n} + \lambda \frac{dP_L}{dP_n} = \lambda \tag{2.48}$$

and with the constraint

$$\sum_{n=1}^{N} P_n - P_r - P_L = 0 \tag{2.49}$$

where

λ = Lagrange multiplier or the cost of received power in dollars per megawatt-hour

[6] K. Denno, "Power System Identification in the Power Flow Reference Frame", *Journal of Applied Science and Engineering A*, Vol. 2 (1977), pp. 141–153.

P_L = the total transmission losses in megawatts
P_r = total received load in megawatts
F_n = the incremental fuel cost of received power in dollars per megawatt
P_n = the output of unit n in megawatts within its maximum capacity

$\frac{dF_n}{dP_n}$ = the incremental production cost in dollars per megawatt

$\frac{2P_L}{2P_n}$ = the incremental transmission losses in megawatts per megawatt

The incremental transmission losses for plant n can be expressed as

$$\frac{dP_L}{dP_n} = \sum_m 2B_{mn}P_m \tag{2.50}$$

where

B_{mn} = the loss formula coefficient
P_m = the output of plant m in megawatts

To calculate the total system losses, the following quadratic equation for the total transmission loss can be used:

$$P_L = \sum_m \sum_n P_m B_{mn} P_n \tag{2.51}$$

The incremental production cost for a given plant can be represented by

$$\frac{dF_n}{dP_n} = F_{nn}P_n + f_n \tag{2.52}$$

where

F_{nn} = the slope of incremental production-cost curve
f_n = the intercept of incremental production-cost curve

Substituting equations 2.50 and 2.52 into 2.48, we get

$$F_{mn}P_n + f_n + \lambda \sum_m 2 \cdot B_{mn}P_n = \lambda \tag{2.53}$$

Therefore,

$$\lambda = \frac{F_{mn} + f_n}{1 - \sum_m 2 \cdot B_{mn}P_n}$$

The procedures of calculating fuel cost after the computation of are as follows. Since λ for each loading of generating units is known, an incremental cost curve (λ versus power output) can be plotted for every generating unit. By integrating such a curve, the fuel-cost curve (input cost versus total power received) can be obtained. The fuel-cost curve provides only the information of fuel cost (dollars per hour) at various loadings. To calculate the total annual fuel cost, the load duration curve (percent power versus percent time) must be introduced. From this curve, the amount of hours for a specific power loading can be secured. Therefore, by using a load duration curve, the fuel-cost curve can be replotted in a fuel cost (dollars per hour) versus total hours (hour) basis. Take the area under such curve; this area represents the total annual fuel cost.

2.2.3 Power Allocation for Storage Batteries and Fuel Cells in a Large-Grid Power System[7]

We are going to present three alternative plans specifying power allocation of storage batteries and fuel cells in a large-grid interconnected power system. This system already had conventional electromechanical power sources supplying initial loads of 10,000 MW.

Plan 1. Fuel cells for dispersed total new load

In this plan, fuel-cell stacks are installed at every load bus bar that requires additional power as well as at every new load center; that is, the entire system will retain the conventional power sources, with all the new demand of power being met by the fuel-cell stack system. The additional load to be supplied by the fuel-cell stacks is about 13,000 MW. Of course, such a system is realized to be dispersed as far as the fuel cell units, which is the ultimate benefit expected from such a pattern of source allocation. Replacing conventional power sources with fuel cells will render environmental benefits as well as reliability and economical gain with the absence of transmission losses.

Plan 2. Storage batteries for peak load

In this plan, all peak loading power occurring throughout the new load of 13,000 MW will be supplied by storage battery plants, each installed in close proximity to the old gas turbine plants. The main advantage of having the storage batteries supply peak load is that they can be charged during off-peak loading duration with lower incremental fuel cost and thus supply peak power demand that would otherwise require higher incremental fuel-cost. Allocation of peak power demand by the storage battery system is designed to reduce the total cost. Of course, this system is very similar to the old systems of fossil steam, nuclear power, gas turbines, and pumped hydroelec-

[7] K. Denno, "Steady-State and Dynamic Investigations . . .".

tric. The total load to be supplied by the expansion of the conventional units and the storage batteries in this plan is about 13,000 MW.

Plan 3. The fuel cell and battery system

In this system, all the steam fossil generators planned for off-peak condition are replaced by the fuel-cell units, and all the gas turbine generators planned for the peak loading condition are replaced by storage batteries. The basic idea is the same as that in plan 2. Plan 3 consists of nuclear, pumped hydro, fuel cell, and storage batteries.

As mentioned earlier, each of three plans carried out additional load of about 13,000 MW above reference plan load supplying about 10,000 MW, the reference plan having its power-plant sources consisting of nuclear pumped hydroelectric for base loading with steam fossil and gas turbine plants supplying off-peak and peak loading, respectively.

2.2.4 The Economics of Fuel Cells and Storage Batteries

To illustrate the economic feasibility of electrochemical power, its integration has actually been carried out in coordination with conventional power sources such as fossil-steam, nuclear, gas turbine, and hydroelectric power plants, as described in plans 1, 2 and 3 earlier.

Economic calculations have been carried out using information presented in Section 2.2.1 and in Table 2.1 in connection with equations 2.48 through 2.52 and in coordination of Figure 2.7 representing a typical load duration curve.

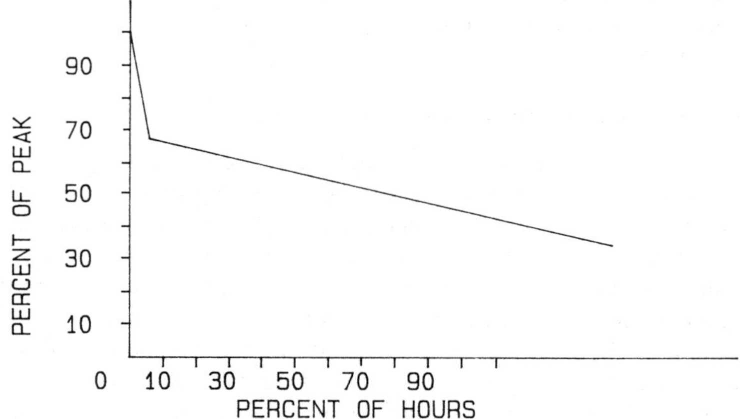

Figure 2.7 Load duration curve. (Reprinted with permission of Energy Utilization of P.S.E. & G., NJ.)

Results of economic computation are shown as follows:

1. Fuel-cost curves for the three plans are as shown in Figures 2.8.a, 2.8.b and 2.8.c.
2. The fuel-cost differential of each plan is given with respect to an all-out base conventional power system. For a conventional system, it is meant that all power supply sources are fossil-cell, gas turbine, nuclear power, and hydroelectric plants. Results for the fuel-cost differential in terms of power demand are shown in Figs. 2.9.a, 2.9.b, and 2.9.c.
3. The fuel-cost differential is given in dollars per hour at any time of the year of each plan with respect to an all-out conventional base power system, as shown in Figs. 2.10.a, 2.10.b, and 2.10.c for plans 1, 2, and 3, respectively.

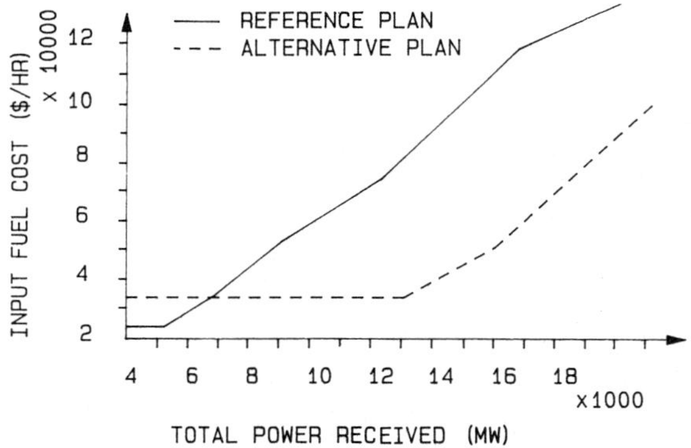

Figure 2.8.a Fuel-cost curve plan 1. (All parts reprinted with permission of Energy Utilization of P.S.E. & G., NJ.)

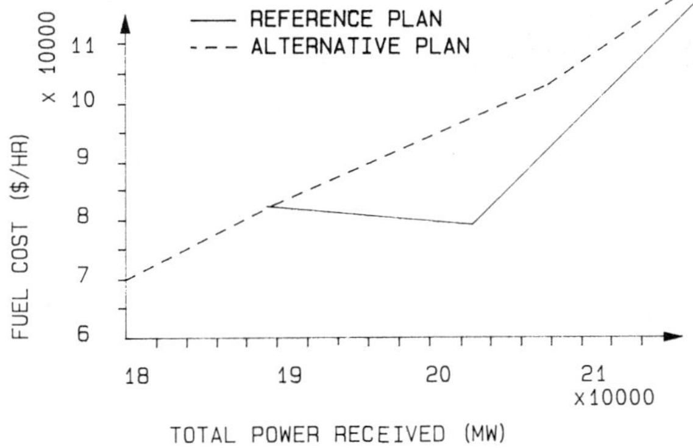

Figure 2.8.b Fuel-cost curve plan 2.

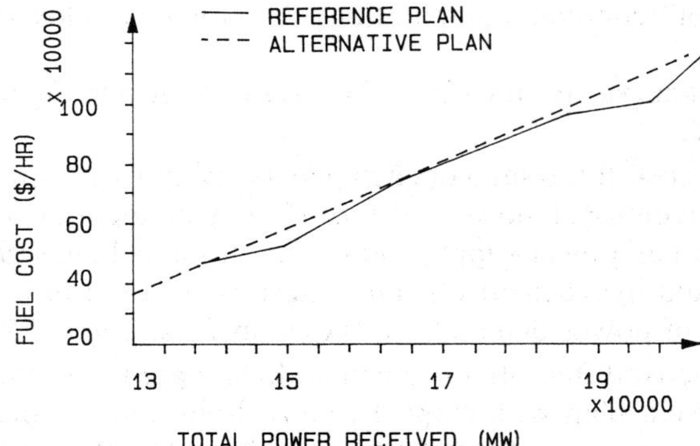

Figure 2.8.c Fuel-cost curve plan 3.

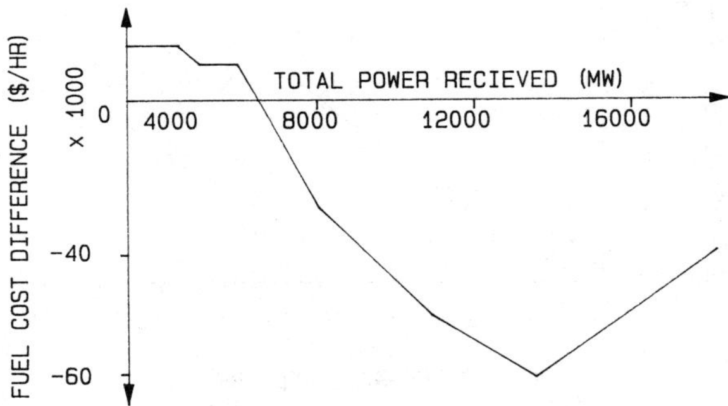

Figure 2.9.a Fuel-cost difference curve plan 1. (All parts reprinted with permission of Energy Utilization of P.S.E. & G., NJ.)

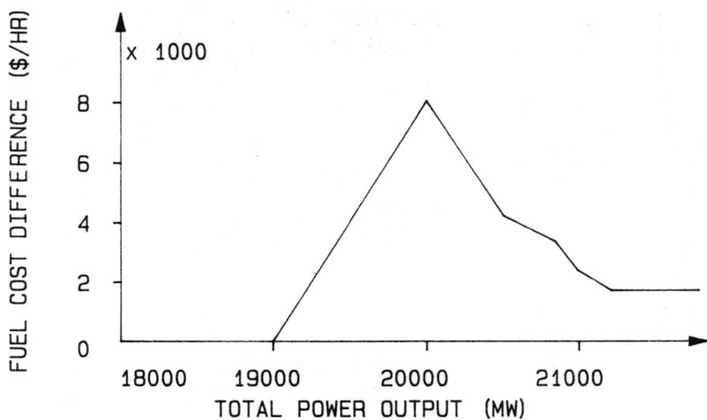

Figure 2.9.b Fuel-cost difference curve plan 2.

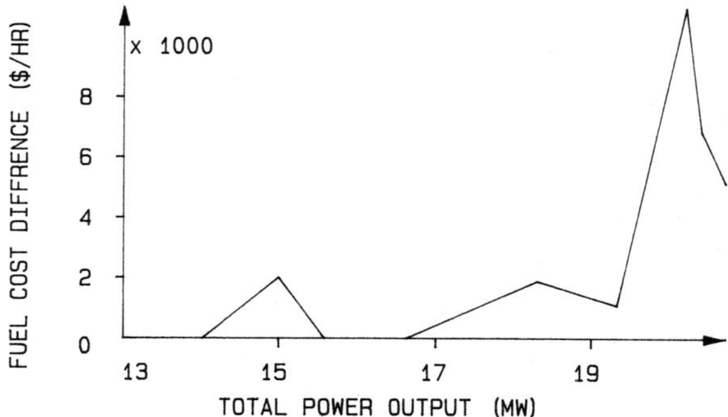

Figure 2.9.c Fuel cost difference curve plan 3.

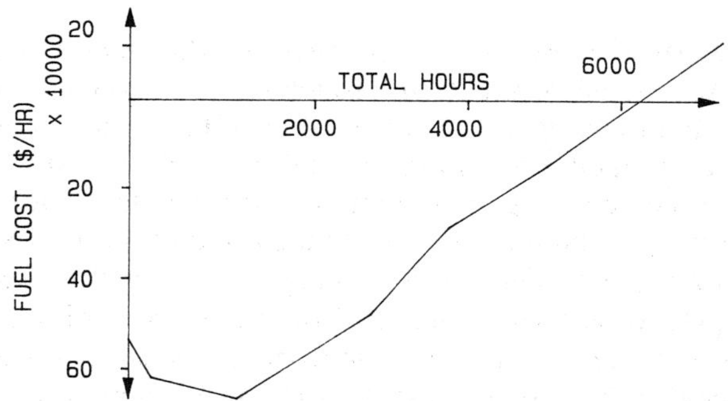

Figure 2.10.a Annual savings curve plan 1. (All parts reprinted with permission of Energy Utilization of P.S.E. & G., NJ.)

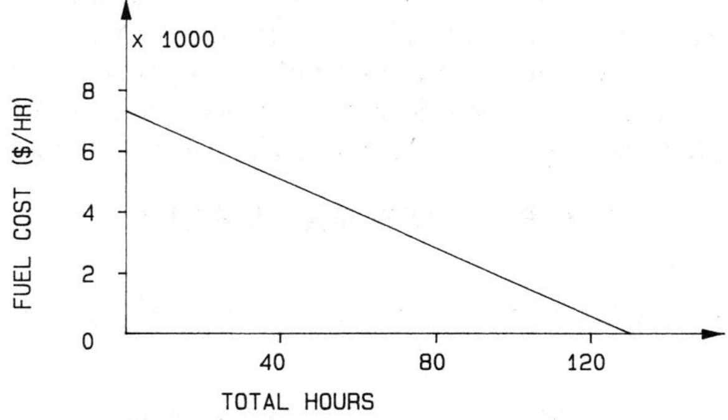

Figure 2.10.b Annual savings curve plan 2.

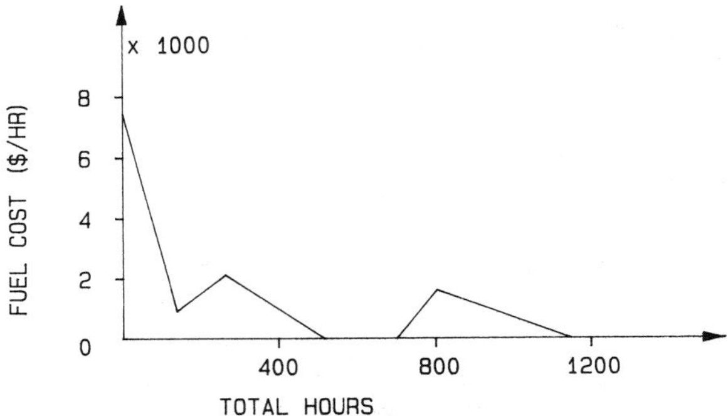

Figure 2.10.c Annual savings curve plan 3.

2.3 SUMMARY

In this chapter, three important areas regarding the electrochemical power sources were presented: simulation, modeling, and economic feasibility.

An equivalent circuit model for the fuel cell and the storage battery in the form of a lumped continuum model was discussed. Mathematical models in the complex frequency domain were developed for the fuel cell and storage battery. And three forms of integrated-grid power systems were presented and analyzed. One form simulated the fuel-cell stack system located in a dispersed pattern, supplying both base and peak loads; a second visualized the storage batteries as supplying peak load replacing the gas turbines in a centralized power system; and a third, which was centralized, utilized fuel-cell stacks to supply off-peak loads (replacing some fossil-steam plants) and storage batteries supplying peak loads, replacing gas turbines.

From the three power-grid systems presented, the second and the third plans were seen as being more economically feasible than the first. However, economic feasibility may change in the future depending upon the cost of any base system and the eventual mode of expansion.

Regarding the concept of environmental impacts, physical realities mentioned indicate clear, favorable environmental benefits of the fuel cell and storage battery as effective and reliable electrochemical power sources.

2.4 SOLVED EXAMPLES

2.4 Solved Examples

1. The equation of electric field decay in a storage battery is expressed by

$$E' = E'_o e^{-t/\tau}$$

If the continuum permittivity is ε, express the total charge residing on either electrode as a function of time and the electric current flow rate.

Solution Electric flux density is $\overline{D} = \overline{E}'$. Therefore,

$$D(t) = E_o' e^{t/\tau} \tag{1.a}$$

With the electric field E perpendicular to the electrode surface (surface A),

$$Q(t) = \varepsilon A E_o' e^{-t/\tau} \tag{1.b}$$

$$i(t) = \frac{dQ}{dt} = -\frac{\varepsilon E_o' A}{\tau} e^{-t/\tau} \tag{1.c}$$

and

$$\frac{di}{dt} = \frac{\varepsilon E_o' A}{\tau^2} e^{-t/\tau} \text{ amp/sec} \tag{1.d}$$

The conclusion that can be drawn now is that a storage battery can be simulated as a lumped-parameter circuit model of a capacitor with an initially stored electrical energy in terms of initial electric potential shunted with a resistance representing the dielectric loss.

2. Given the simple mathematical model for the DC storage battery expressed by equation 2.4, solve for the time response of $E_d(t)$.

Solution

$$E_d(s) = \frac{E_{do}\tau}{1 + \tau S} \tag{2.a}$$

where E_{do} is the initial potential difference stored in the battery.

Taking the inverse Laplace transform of equation 2.a results in

$$E_d(t) = E_{do}\, \mathcal{L} \frac{1}{S + 1/\tau} \tag{2.b}$$

$$= E_{do} e^{-t/\tau} \tag{2.c}$$

3. Refer to equation 2.26 for the electrochemical reaction at the cathode of hydrogen-oxygen fuel cell. Calculate the heat release in joules per kilogram-mole if the cell emf generated is 1 volt at standard pressure and temperature, the enthalpy of energy formation for liquid water is -286×10^6 joules/kg-mole, and the universal gas constant $R = 8\,314$ joules/kg-mole-K.

Solution From equation 2.27,

$$2H_2 + O_2 \rightarrow 2H_2O + 4eE + Q - 3RT$$

In terms of energy enthalpy formation,

$$O_2 = -2 \times 286 \times 10^6 + 4 \times 1.6 \times 10^{-19} - 3 \times 8314 \times 298 + Q$$

where $Q = 579$ joules/kg-mole.

4. Given

$$\Delta E = \frac{|\Delta G| - |We|}{nF} \tag{2.4.1}$$

where

$$\Delta E = \text{the perturbation in the fuel-cell emf in volts}$$
$$W_e = NFE$$

If ΔG and n are constants at standard pressure and temperature, solve for the normalized differential perturbation in E; that is,

$$\frac{\partial E}{E}.$$

Solution From equation 2.4.1,

$$E = \frac{|\Delta G|}{nF} - \frac{NFE}{n}$$

$$= \frac{\Delta G}{nF} - \frac{NE}{n} \tag{4.a}$$

or, replacing Δ with ∂ for differential perturbation,

$$nF\partial E = \partial G - NFE$$

$$nF = \frac{\partial E}{E} = \frac{\partial G}{E} - NF$$

$$\frac{1}{E}[nF\ E - G] = -NF \tag{4.b}$$

but $\partial G \geq EnF$. Therefore, from 4.b,

$$\frac{1}{E}[nF\partial E - \geq EnF] = -NF \tag{4.c}$$

Hence, we can see from 4.c, that

$$n \geq -N = \frac{n\partial E}{E} \tag{4.d}$$

or

$$\frac{\partial E}{E} \geq \frac{N}{n} - 1 \tag{4.e}$$

Equation 4.d implies that a normalized perturbation in E is directly measured by the equivalent orders of electrochemical reactions per mole or reactants less unity.

5. Refer to the process of fuel-cell modeling expressed by equations 2.32–2.34. Find the required relationship for electrochemical potential of a reactant in terms of molar concentration, equivalency of chemical reaction, and the cell emf. Assume $dE = 0$; that is, the cell emf is steady state and without any perturbation.

Solution From equations 2.30 and 2.33, with $dE = 0$,

$$dG = \mu dn \tag{5.a}$$

$$dG = FEdn + FEdn \tag{5.b}$$

Therefore,

$$dn = FE(dn + dN) \tag{5.c}$$

or

$$\mu = FE\left(1 + \frac{dN}{dn}\right) \tag{5.d}$$

Equation 5.d states that the electrochemical potential of a reactant is a linear function in terms of the rate of change of an ordered chemical reaction equivalency function with respect to molar concentration.

6. Given:

$$F_1 = (8P_1 + 0.018P_1^2 + 60)10^6 \tag{6.a}$$

$$F_2 = (4P_2 + 0.03P_2^2 + 100)10^6 \tag{6.b}$$

Assume no transmission loss. Express the cost of generation in terms of P_1 and P_2.

Solution From equation 2.48, and since $P_L = 0$,

$$P_1 = \frac{dF_1}{dP_1} = (8 + 0.036)10^6 \tag{6.c}$$

and

$$P_2 = \frac{dF_2}{dP_2} = (4 + 0.06P_L)10^6 \tag{6.d}$$

In a metropolitan system like the one used in this study, the effect of the transmission losses is neglected in the scheduling of generation, so the optimum loading can be expressed in the following way by assuming B_{mn} to be zero:

$$\lambda = P_nF_{nn} + f_n \tag{6.e}$$

because

$$\frac{dF_n}{dP_n} = F_{nn}P_n + f_n \tag{6.f}$$

therefore,

$$\lambda = \frac{dF_n}{dP_n} \tag{6.g}$$

We have to keep in mind that equation 6.e is true only in a metropolitan system where losses are assumed to be negligible ($P_L = 0$).

The meaning of equation 6.f is very important for the following reasons:

When equation 6.*f* is satisfied, total capital input (F_t) to a system is at its minimum. However, total output power (P_n) is at its optimum. In other words, the minimum input (dollars per hour) for a given load is obtained when all generating units are operated at the same incremental cost (λ) found in equation 6.g. Any change in λ will cause the same change in total generation as indicated in the equation. Therefore, if λ is known for any generating capacity, one can be sure that the most economic scheduling of generation is obtained. The fuel cost computed from this λ will be the most economic fuel cost.

7. Referring to Fig. 2.8.a, establish relationships for the cost of fuel input versus total power ($P \geq 14{,}000$ WM) for the reference plan and alternative plan where the fuel cells have been allocated in a dispersed mode of power supply.

Solution For $P \geq 14{,}000$ MW, cost of fuel input follows a straight-line pattern. Therefore,

FUEL-CELL PLAN

$$P_1 = 14{,}000 \text{ MW}, \ F_1 = \$11{,}750/\text{hr}$$

$$P_1 = 18{,}000 \text{ MW}, \ F_1 = \$13{,}000/\text{hr}$$

REFERENCE SYSTEM

$$P_2 = 14{,}000 \text{ MW}, \ F_2 = \$4000/\text{hr}$$

$$P_2 = 18{,}000 \text{ MW}, \ F_2 = \$7300/\text{hr}$$

FUEL CELL SYSTEM

$$F_1 = aP_1 + b \tag{7.a}$$

Hence

$$11{,}750 = a14{,}000 + b$$

$$13{,}000 = a18{,}000 + b$$

Therefore,

$$4000a = 1250$$

$$a = \frac{125}{400} - \frac{5}{16} \tag{7.b}$$

and

$$b = 7375 \tag{7.c}$$

Hence,

$$F_1 = \frac{5}{16} P_1 + 7375 \tag{7.d}$$

Similarly, for the reference system,

$$F_2 = 0.825P_1 - 7155$$

8. Refer to Fig. 2.10.b and calculate the dollar savings for plan 2 for having storage batteries supply peak load.

Solution Figure 2.10.b shows a straight-line pattern for annual savings in dollars per hour versus total hours. Hence, total savings annually is equal to the area A under the curve. Therefore,

$$A = \frac{78{,}000 \times 130}{2} = \$507{,}000$$

Now, let us offer some tangible comments regarding the economic picture presented in Figures 2.8, 2.9, and 2.10.

First, with respect to Figures 2.8.a, 2.8.b, and 2.8.c showing fuel costs in dollars per hour versus power demand, we can see that plan 1 illustrated in Figure 2.8.a indicates a costly level with respect to a totally conventional grid system, while plan 2 demonstrates less economic differential and plan 3 shows an almost even economic level with respect to a totally conventional system.

Second, the total fuel-cost differentials versus load power demand for each of the three plans mentioned earlier are shown topographically in Figures 2.9.a, 2.9.b, and 2.9.c. The economics are favorable for plans 2 and 3.

Third, turning to the total time–dollar savings at any time of the year, we can see the sizable dollar loss for plan 1, where fuel-cell units are located in a dispersed system of allocation to supply a load of about 13,000 MW covering base as well as peak loads. However, in plans 2 and 3, economic gain could be achieved at some time duration regarding the supply of peak load by the storage batteries as in plan 2 and with the fuel cells supplying the off-peak load and storage batteries the peak load as in plan 3.

2.5 REVIEW QUESTIONS

1. State reasoning behind the fact that the fuel cell may follow the laws of thermodynamics, only close to their boundaries.
2. Explain the circuit concept for the storage battery as composed of a capacitance in shunt with a high resistance.
3. Regarding the lumped-circuit representation of the storage battery, if the ε of the battery continuum is doubled, while its electrical conductivity is tripled, find the effect on the state of charge relaxation.
4. Comment on the statement that the equivalent order number (N) of an electrochemical reaction is not an integer.
5. Explain the reasoning that Gibbs free energy function is $\Delta G \gg nFe$, where F is Faraday's constant.
6. Discuss and comment on the life aspect of fuel cells and storage batteries in conjunction with maintenance and operation costs with respect to conventional power sources.
7. Refer to equation 2.48; discuss the economic implications of the cost of power generation and the cost of transmission loss.
8. Derive equation 2.53, from your knowledge of F_n and P_L.
9. Tabulate the favorable and unfavorable environmental impacts of the fuel cell and storage acid battery.

10. Sketch a block diagram showing systematic connection of the fuel cell and storage battery to supply an AC output power.

2.6 PROBLEMS

1. Derive an equation describing the state of relaxation for the electric charge density in the storage battery three-dimensional continuum where

$$\overline{\nabla} \times \overline{E} = ay\, U_o(z)B$$
$$U_o(X) = \text{a unity impulse}$$
$$B = \text{the impulse amplitude}$$

Consider that E_x is a function of Z and $E_y = E_Z = 0$.

2. Establish a model for the storage battery in the complex frequency domain based on the result of Problem 1.
3. State the first and second laws of thermodynamics and present analytical reasoning of why the fuel cell may follow them with some justification.
4. In reference to the Section 2.1.7 if

$$n(s) = \frac{k_1}{S}$$

$$N(s) = \frac{1 + S}{k + S^2}$$

$$Q(S) = \frac{S}{k_1 + 4S^2}$$

then

$$k_1 = \text{a constant}$$

Modify equation 3 with $G' = G'' = 0$.

5. Given that

$$F_1(P_1) = 0.0015P_1^2 + 0.31P_1 + 7375$$

$$F_2(P_2) = 0.0020P_2^2 + 0.825P_2 - 7155$$

$$\text{total received load} = 22{,}000 \text{ MW}$$

$$P_{1_{max}} \leq 10{,}000 \text{ MW}$$

$$P_{2_{max}} \leq 12{,}000 \text{ MW}$$

$$P_L = 0$$

Solve for the cost of received power in dollars per megawatt hour and optimum power allocation to P_1 and P_2.

6. Repeat solution of Problem 5 with

$$P_L = 0.0003P_1^2 + 0.0001P_2^2$$

7. Establish from the principle of piecewise linearization a ratio of nondimensional equation for the fuel cost in dollars per hour of a system using storage batteries supplying peak load to a totally conventional power system. Refer to Fig. 2.8.b and calculate the ratio for $p = 20{,}000$ MW.
8. Repeat Problem 7, but with respect to Fig. 2.8.c.
9. Refer to Fig. 2.10.a and calculate the total dollar loss or savings for plan 1, involving the fuel cells supplying power in a dispersed system.
10. Repeat Problem 9, but for Fig. 2.10.c, for a system having fuel cells supplying off-peak load and storage batteries supplying peak load.
11. Solve for the current transient response of a storage battery supplying steady load when suddenly subjected to a short circuit.
12. Repeat Problem 11, but for a conventional hydrogen fuel cell supplying steady variable loading conditions.
13. Establish a unified model in the complex frequency domain for a fuel cell linked to a storage battery. (Assume $G' = G'' = 0$ in the fuel-cell model.)
14. Secure the time response of the model established in Problem 13, when a sudden short circuit occurred at the terminals of the fuel cell.
15. Repeat Problem 14 with a sudden increase in the total resistive load connected across the combination of the storage battery and fuel-cell set.
16. Repeat Problem 1, but with cyclic variations along the Z and y orientations and where

$$x\overline{E} = a_y U_o(Z)B + a_z U_o(Y)B \quad \text{and} \quad E_Z = 0$$

The battery continuum is three dimensional. Consider E_x as a function of Z and Y. Both a_y and a are unit vectors in the Y and Z axes, respectively.
17. A storage battery supplies a resistive load R_2 in the steady state, when a sinusoidal magnetic expressed by $B(t) = a_z B_{\max} \sin \omega t$ is suddenly subjected on the X-Y plane containing the battery and the load in a closed loop. Solve for the resultant current flow to the load, including the transient and steady state. You may assume the loop has effective cross-sectional area A and electrical conductivity of σ.
18. Repeat Problem 17 for the case of a fuel cell supplying a load R_L.
19. Consider a power system involving the series linkage of a fuel cell and storage battery supplying a load R_L, when suddenly the load is reduced by 50 percent. Solve for the time response of the load current $i_L(t)$.
20. Establish a mathematical model in the complex frequency domain for a fuel-cell series-linked power assembly connected across a DC shunt motor. Find the time response of the line current during the starting period. You may assume appropriate initial conditions.
21. Establish a mathematical model in the complex frequency domain for a set of series interconnected storage batteries across a series DC motor under the mechanical load. Express the time response of the line current due to a sudden reduction in the mechanical torque at the motor.
22. During battery discharge or recharge, the closed loop with the load or supply, an inductive effect does exist. Establish a mathematical model in the complex fre-

quency domain and then deduce the physically realizable lumped-parameter-equivalent network.

2.7 REFERENCES

1. Crouch, D. A., and J. Werth. "Batteries and the Economics of Load Leveling in Large Power Plants." Paper presented at the winter 1973 meeting of the Power Engineering Society of IEEE, Jan./Feb. New York.
2. Denno, K. "Steady-State and Dynamic Investigations for Determining Optimum Electrochemical Electromechanical Inter-Connected Power Systems." Published report for a research grant supported by the Middle Atlantic Power Research Committee, January 1985.
3. Denno, K. "Economic Optimization for an Integrated Power System" (with C. Mollo and W. Yang). *Midwest Power Symposium Proceedings,* Rolla, Missouri Oct. 1974.
4. Denno, K. "Steady-State and Dynamic Modeling of an Integrated Power System." *Proceedings of Canadian Communication and Power Conference,* Montreal, November 1974.
5. Denno, K. "Power System Identification in the Power Flow Reference Frame." *Journal of Applied Science and Engineering A,* Vol. 2 (1977), pp. 141–153.
6. Denno, K. "Compatibility of Direct Energy Storage Devices with AC Processing Power Systems Components." *Proceedings of Energy 78 Conference,* pp. 6–10, April 1978 in Tulsa, Oklahoma.
7. Denno, K. "Mathematical Modeling of Storage Battery and Fuel Cell." *Proceedings of the 1978 International Telephone Energy Conference,* pp. 237–243, Oct. 1978 in Washington, D.C.
8. Eklund, L. G., and W. S. Lueckel. "Fuel Cells for Dispersed Power Generation." *IEEE Transaction on Power Apparatus and Systems,* Vol. 92, no. 1 (January/February 1973), pp. 230–236.
9. Heredy, L. A., and W. E. Parkins. "Lithium-Sulfur Battery Plant for Power Peaking." Paper presented at the Winter 1972 meeting of the Power Engineering Society of IEEE.
10. Liebhafsky, H. A. and Cairns, E. J. *Fuel Cells and Fuel Batteries,* New York: John Wiley & Sons, Inc., 1968.
11. Mantell, C. *Batteries and Energy Systems.* New York: McGraw-Hill Book Co., 1970.
12. Mollo, C. R., and K. Denno. "The Feasibility of Fuel Cells in Modern Power Systems." Master thesis, NJIT, Newark, N.J. 1973.
13. Yang, W. K., and K. Denno. "The Feasibility of Fuel Cell in a Modern Power System." Master thesis, NJIT, Newark, N.J. 1973.

3

Bioelectrochemical Conversion of Refuse to Synthetic Fuel and Energy

3.1 INTRODUCTION

Disposal of organic wastes from refuse is being redirected toward the production of synthetic bioorganic fuels, on one hand (already being used as supplemental fuel for large power-plant boilers), and the generation of electric energy, on the other. Statistical information indicates that from dry, ash-free organic waste on the order of 880 million tons, energy equivalent to about 170 million barrels of oil, or 1.36 trillion standard cubic feet of methane, could be obtained. Procedures for extracting synthetic bioorganic fuels from refuse may be based on the processes of hydrogeneration pyrolysis and biochemical conversion.

The electrochemical output of a biochemical fuel cell is related to the rate of bacterial metabolism; that is, the result of this process of combustion by the microorganisms is the electric power output. Current literature indicates that if the rate of bacterial nutrition could be high, operational performance of the biochemical fuel could be close to that of regular fuel cells. However, lacking knowledge of the kinetics and dynamics of the biochemical reactions, the present level of electric power density output of the biochemical cell remains relatively low.

Research efforts to extract bioorganic fuel C_mH_{2m+2} (where m is the order of the hydrocarbon) from refuse and the direct utilization of the bioorganic fuel in the performance of the power cell has to be increased theoretically and experimentally, especially when considering the use of worthless materials, such as sewage, garbage, grasses, algae, and sawdust, are available in ample amounts.

This chapter describes the processes of synthetic fuel extraction and the results of direct use of the bioorganic fuel in a power cell. The results obtained, which are theoretical, are secured through the interaction of material reactants with the local and external environments via mathematical and dynamic modeling.

Given the availability of a prepared bioorganic fuel CH_2O, together with all favorable conditions for the interactions of material reactants with the environment, it is necessary to:

1. Develop mathematical and dynamic models for the total operational performance of the biochemical fuel cell and the bioconversion extraction of hydrocarbon fuel from prepared refuse.
2. Identify the effective parametric role of the prepared bioorganic refuse, such as activity, concentration, entropy, and enthalpy, on the responsive ratio of energy content with respect to concentration of reactants, in the process of energy extraction.
3. For biochemical fuel cells, identify the effective roles of the bioorganic fuel, type of electrolyte, sources of internal voltage polarizations, and rate of oxidization on the responsive ratio of output terminal voltage with respect to energy content of bioorganic fuel.

3.2 BIOFUEL AS PERFECT ELECTROCHEMICAL APPARATUS

We may look at the biochemical fuel cell as an electrochemical power generator in which the bioorganic matter serves as the fuel source. A reasonably good biofuel cell involves the catalytic oxidation by preheated air at the cathode of the bioorganic fuel released by the anode by the action of microorganisms. Electrolyte solution is assumed to be potassium hydroxide. The biofuel cell reaction can therefore be expressed as follows:[1]

For the anode reaction,

$$\underset{\text{(biofuel)}}{\tfrac{1}{2}CH_2O} + 2OH \rightarrow \tfrac{1}{2}CO_2 + \tfrac{3}{2}H_2O + 2e \tag{3.1}$$

For the cathode reaction,

$$\tfrac{1}{2}O_2 + H_2O \rightarrow 2OH - 2e \tag{3.2}$$

For the total electrochemical reaction,

$$\tfrac{1}{2}CH_2O + \tfrac{1}{2}O_2 \rightarrow \tfrac{1}{2}CO_2 + \tfrac{1}{2}H_2O \tag{3.3}$$

[1] © 1963 IEEE. Reprinted, with permission, from Proceedings of the IEEE, Vol. 51, No. 5, May 1963, pp. 812–819. Paper entitled: "Preliminary Biochemical Fuel Cell Investigation" by E. L. Colichman.

Present operational performance of biofuel cells without the auxiliary addition of bacteria, but with 40 percent KOH, gives peak power density of the order of 2 mW/cm^2 at an output terminal voltage of around 0.3 volt and current density of 8 ma/cm^2.

Perfection of the local electrochemical reaction is based on the assumption that no significant internal changes take place in the cell except that all variations that are associated with the flow of current are reversible.

There are two requirements for a perfect biocell: reactivity and invariance.

REACTIVITY REQUIREMENTS

1. *The stoichiometric condition.*[2] To obtain maximum amount of electricity from oxidization of 1 mole of bioorganic fuel, H_2O and CO_2 are the most probable products of reaction since they represent the oxidization product of lowest energy content.
2. *Kinetic condition.* This implies smooth coupling of reactants with high electrode activity. The rate of release of valence electrons to load depends upon the rate of reaction at the electrodes. Consequently, this will establish a higher level of terminal voltage.

INVARIANCE REQUIREMENTS

1. No corrosion of side reactions
2. Invariant electrolyte
3. No serious variation in electrodes composition

However, biochemical reaction in the biofuel cell practically will generate products other than CO_2 and H_2O, leading to the conclusion that the order of total effective electrochemical reaction is not an integer in nature.

Nevertheless, practical operation of the biofuel cell is in fact irreversible, rendering the definite existence of voltage change due to actual difference between Gibbs free energy and the reversible electrical output as expressed by the relation

$$\frac{|G| - |W_e|}{nF} = \Delta E \tag{3.4}$$

where

W_e = reversible electrical energy output in joules = NFE

F = the expected quantity of electricity represented by 1 kg-mole of electrons and is equal to 96.5×10^6 coulomb

[2] K. Denno, "Bio-electrochemical Conversion of Refuse to Energy," presented at the 1979 Annual Technical Meeting, Institute of Environmental Sciences, pp. 316–321. Also from H. A. Liebhafsky and E. J. Cairns, *Fuel Cells and Fuel Batteries* (New York: John Wiley & Sons, Inc. 1968).

N = the order of equivalent biochemical reactions, which is most likely not an integer
n = equivalent molar concentration of reactants = $n_1 + n_2$
n_1 = molar concentration of biofuel
n_2 = molar concentration of oxygen
ΔE = change in the biocell terminal voltage from no load
ΔG = Gibbs free energy, defined as the maximum amount available at constant pressure and temperature = $\Delta H - T\Delta S$
ΔH = work exchange with the atmosphere in terms of enthalpy
$T\Delta$ = work exchange with the local environment of the biocell

3.3 KINETICS AND DYNAMICS OF THE BIOELECTROCHEMICAL CELL[3]

ΔG is considered as the maximum energy available out of a biochemical reaction at constant pressure and temperature. And it can be expressed as the sum of all individual biochemical potentials of all reactants involved:

$$G = \sum_{i=1}^{i=j} \mu_i dn_i \tag{3.5}$$

where μ_i is the electrochemical potential at the ith reactant. The overall energy equation is

$$dU = TdS - dG \tag{3.6}$$

where

dU = total energy differential

dS = total entropy differential

Irreversibility is an actual operational feature of the biofuel cell where

$$nFE = G - NFE \tag{3.7}$$

or

$$dG = F[ndE + Edn] + F[Nde + EdN] \tag{3.8}$$

Then imposing the state of time variation and neglecting volume change within the biocell continuum, equations 3.5 and 3.7 become

$$\frac{dU}{dt} = F\left(n\frac{dE}{dt} + E\frac{dn}{dt}\right) + F\left(N\frac{dE}{dt} + E\frac{dn}{dt}\right) + T\frac{d}{dt} \tag{3.9}$$

[3] Ibid.

To proceed now for the development of a dynamic model for the bioelectrochemical fuel cell, the following basic rules of Laplace transformation must be cited:

$$[f_1(t)f_2(t)] = \sum_{k=1}^{P} \frac{A_1(S_k)}{B_1'(S_k)} F_2(S - S_k) \tag{3.10}$$

where $A_1(S_k)/B_1(S_k)$ is the rational transform function for $f_1(t)$ and S_k are the poles in $B_1(S_k)$.

In equation (3.8), $f_1(t)$ is $n(t)$ and $f_2(t)$ is $N(t)$. Therefore,

$$\begin{aligned} n(S) &= n_1(S) + n_2(S) = \frac{A_1(S_{k_1})}{B_1(S_{k_2})} \\ N(S) &= N_1(S) + N_2(S) = \frac{A_2(S_{k_2})}{B_2(S_{k_2})} \\ Q(S) &= Q_1(S) + Q_2(S) = \frac{A_3(S_{k_3})}{B_3(S_{k_3})} \end{aligned} \tag{3.11}$$

Subscripts 1 and 2 refer to the bioorganic fuel and preheated air as the two major reactants. Note also that

$$E(S) = \frac{E}{S} \tag{3.12}$$

Then using equation 3.10, the following transform equation is secured with

$$\frac{E(S)}{U(S)} = G - \frac{U_o/S}{U(S)} G + \frac{E_o/S}{U(S)} G' - \frac{1/FS}{U(S)} G'' \tag{3.13}$$

where

$$G = \frac{1/S[n(s) + N(S)]}{1 + \dfrac{\sum_{k_1=1}^{P_1} [A_1(S_{k_1})/B_1'(S_{k_1})] + \sum_{k_2=1}^{P_2} [A_1(S_{k_2})/B_2'(S_{k_2})]}{S[n(s) + N(S)]}} \tag{3.14}$$

$$= \frac{G_1}{1 + G_1 H_1}$$

$$G_1 = \frac{1}{F} \frac{1}{S[n(s) + N(S)]} \tag{3.15}$$

$$H_1 = F\left[\sum_{k_1}^{P_1} \frac{A_1(S_{k_1})}{B_1'(S_{k_1})} + \sum_{k_2}^{P_2} \frac{A_2(S_{k_2})}{B_2'(S_{k_2})}\right] \tag{3.16}$$

then

$$G' = \frac{G_1'}{1 + G_1'H_1'} \tag{3.17}$$

$$G' = \frac{\dfrac{\sum_{k_1}^{P_1} [A_1(S_{k_1})/B_1'(S_{k_1})] + \sum_{k_2}^{P_2} [A_2(S_{k_2})/B_2'(S_{k_2})]}{S[n(s) + N(S)]}}{1 + \dfrac{\sum_{k_1}^{P_1} [A_1(S_{k_1})/B_1'(S_{k_1})] + \sum_{k_2}^{P_2} [A_2(S_{k_2})/B_2'(S_{k_2})]}{S[n(s) + N(S)]}} \tag{3.18}$$

where

$$G' = \frac{\sum_{k_1}^{P_1} [A_1(S_{k_1})/B_1'(S_{k_1})] + \sum_{k_2}^{P_2} [A_2(S_{k_2})/B_2'(S_{k_2})]}{S[n(s) + N(S)]} \tag{3.19}$$

$$H_1' = 1 \tag{3.20}$$

and

$$G'' = \frac{\dfrac{\sum_{k_3}^{P_3} [A_3(S_{k_3})/B_3'(S_{k_3})]\, T(S - S_{k_3})}{\dfrac{\sum_{k_1}^{P_1} [A_1(S_{k_1})/B_1'(S_{k_1})] + \sum_{k_2}^{P_2} [A_2(S_{k_2})/B_2'(S_{k_2})]}{S[n(s) + N(S)]}}}{\sum_{k_1}^{P_1} [A_1(S_{k_1})/B_1'(S_{k_1})] + \sum_{k_2}^{P_2} [A_2(S_{k_2})/B_2'(S_{k_2})]} \tag{3.21}$$

where

$$G_1'' = \frac{\sum_{k_3}^{P_3} [A_3(S_{k_2})/B_3'(S_{k_3})]\, T(S - S_{k_3})}{\sum_{k_1}^{P_1} [A_1(S_{k_1})/B_1'(S_{k_1})] + \sum_{k_2}^{P_2} [A_2(S_{k_2})/B_2'(S_{k_2})]} \tag{3.22}$$

$$H_1'' = S[n(s) + N(S)] \sum_{k_3}^{P_3} [A_3(S_{k_3})/B_3'(S_{k_3})]\, T(S - S_{k_3}) \tag{3.23}$$

The dynamic model for the biofuel cell based on equation 3.13 is shown in Fig. 3.1. This model relates the electrical output potential with respect to the total internal energy content of the bioorganic fuel in terms of concentration of reactants, total entropy, temperature gradient, and irreversible order of bioelectrochemical reactions.

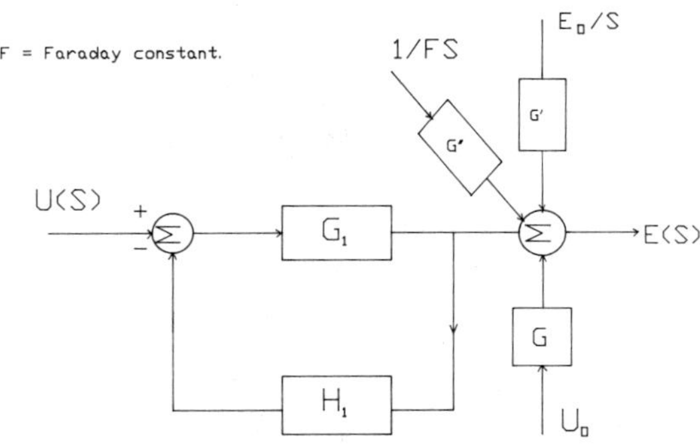

Figure 3.1 Dynamic model for the biofuel cell. (Presented at the 1979 Annual Technical Meeting, Institute of Environmental Sciences and published in the Proceedings. Reprinted with permission.)

3.4 DYNAMICS OF BIOCONVERSION

This process involves the extraction of multigrades of saturated and unsaturated hydrocarbons on the order of C_mH_{2m+2} (m is the order of the hydrocarbon) from refuse. This process is again assumed to occur through the interaction of material reactants, namely, the refuse with the oxidizing environment.

Assume a certain quantity of prepared refuse at temperature T_i, pressure P_i, and concentration n_i to interact with a given oxidizing environment containing unlimited substances at temperature T_o, pressure P_o, and concentration n_o.

Then we let W be defined as the maximum amount of energy that can be obtained from the prepared refuse out of its interaction with the environment continuum:[4]

$$W = \sum_i n_i W_i \tag{3.24}$$

where subscript i refers to any part of the prepared multicompound refuse. We may now write

$$w_i = \frac{\partial W}{\partial n_i} \tag{3.25}$$

and

[4] Reprinted with permission from *Energy Conversion,* Vol. 15, pp. 81–84, Lothar Reikert, "The Conversion of Energy in Chemical Reactions," © 1976, Pergamon Press, Ltd.

$$W_i = W_{ic} + W_{i\nu} \tag{3.26}$$

$$W = W_c + W_\nu \tag{3.27}$$

$$W_{ic} = \frac{\partial W_{ic}}{\partial n_i} \tag{3.28}$$

$$W_{ipT} = \frac{\partial W_{i\nu}}{\partial n_i} \tag{3.29}$$

where

$W_{i\nu}$ = maximum amount of equivalent work extracted per mole due to a change in volume for the ith reactant

W_{ic} = maximum amount of equivalent work extracted per mole due to a change in chemical composition of the ith reactant

W_{ipT} = maximum amount of equivalent work extracted per mole at variable pressure and temperature known as energy

Also from equations 3.24, 3.28, and 3.29 we can observe the following implication:

$$W = W_{PT} + W_c \tag{3.30}$$

And we may cite from the basic laws of thermodynamics, namely, that[5]

$$W_{PT} = H - H_o - T_o(S - S_o) \tag{3.31}$$

$$W_C = \sum_i n_i(W_{ic}^o + RT_o \ln A_i) \tag{3.32}$$

where

S = the net entropy content of the reaction process

A_i = the molecular activity of the ith multicompound reactant

R = the universal gas constant, 8.31 joule/mole/K

W_{ic}^o = the equivalent work extraction per mole of pure substance

Then we may be considering the fact that the order (m) of the produced hydrocarbon in C_mH_{2m+2} is proportional inversely with the activity A,

$$A_i = \frac{M}{n} \tag{3.33}$$

where M is a constant that is usually identified as the value of m at the initial place and time of reaction. So equation 3.32 becomes[6]

$$W_c = \sum_i n_i \left(W_{ic}^o + RT_o \ln \frac{M}{m}\right) \tag{3.34}$$

[5] Ibid.

[6] K. Denno, "Bio-electrochemical Conversion of Refuse to Energy," presented at the 1979 Annual Technical Meeting, Institute of Environmental Sciences, pp. 316–321.

and equation 3.30 can be written as

$$W = H - H_o - T_o(Q - Q_o) + \sum_i n_i \left(W_{ic}^o + RT_o \ln \frac{M}{m}\right) \quad (3.35)$$

or

$$\frac{\partial W}{\partial t} = \frac{\partial H}{\partial t} - T_o \frac{\partial Q}{\partial t} + n_i \frac{\partial W_{ic}^o}{\partial t} + \left(W_{ic}^o + RT \ln \frac{M}{m}\right) \frac{\partial n_i}{\partial t} \quad (3.36)$$

with m now assumed to be independent of time.

Taking the Laplace transforms of equation 3.36 and using the same rules as in Section 3.3, the following form in the complex frequency domain is obtained:

$$\frac{W(S)}{n(S)} = \frac{S}{n(S)} [H(S) - T_oQ(S)] + SRT_o \ln \frac{M}{m} + \sum_{k_1}^{P_1} \frac{C_1(S_{k_1})}{D_1'(S_{k_1})} \frac{C_2(S - S_{k_1})}{D_2(S - S_{k_1})} (2S - S_k) \quad (3.37)$$

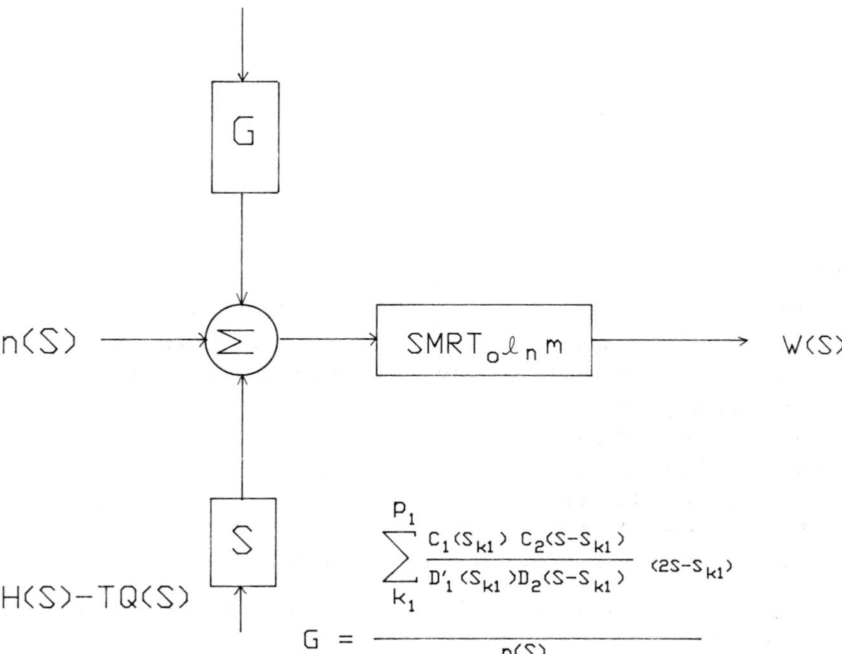

Figure 3.2 Dynamic model for bioconversion of prepared refuse to organic fuel. (Presented at the 1979 Annual Technical Meeting, Institute of Environmental Sciences and published in the Proceedings. Reprinted with permission.)

where

$$W_{ic}^{o}(S) = \frac{C_1(S)}{D_1(S)} \tag{3.38}$$

$$n(S) = \frac{C_2(S)}{D_2(S)} \tag{3.39}$$

and S_k are poles of $C_1(S)/D_1(S)$. From equation 3.37 a dynamic model is developed as shown in Fig. 3.2.

3.5 CONVERSION OF HYDROCARBON TO ELECTRIC ENERGY[7]

Let us resume our analysis from the following set of energy and continuity equations expressed earlier:

$$U = TS - FnE - FNE \tag{3.40}$$

$$G = NFE - FnE = FE(N - n) \tag{3.41}$$

$$= \sum_{i=1}^{j} \mu_i n_i \tag{3.42}$$

$$\begin{aligned} \frac{\partial n_i}{\partial t} &= \sum_i Y_i - n_i \frac{\partial \nu_i}{\partial x} - \nu_i \frac{\partial n_i}{\partial x} \\ &= \sum_{i=1}^{n} Y_i - \overline{\nabla} \cdot n_i \overline{V}_i \end{aligned} \tag{3.43}$$

where

U = the total energy content of the entire reaction process
Y_i = the time rate of change of concentration of the ith reactant
V_i = the local average fluid velocity of synthetic fuel continuum
G = Gibbs energy function
F = Faraday's constant
N_i = electrochemical reaction order of the ith reactant
E = electric potential output of the biocell
n_i = concentration of the ith reactant
μ_i = equivalent potential of electrochemical reaction

The constraint criterion developed for maximum energy transfer in space and time domain is

[7] K. Denno, "Simulating Criterion for Biochemical Conversion of Refuse to Synthetic Fuel and Electric Power," *Alternative Energy Sources III*, Vol. 7, (1980) pp. 241–247, Hemisphere Publishing Corp.

$$\frac{T}{F}\left(\frac{\partial}{\partial t} - \frac{\nu_i}{E}\frac{\partial}{\partial x}\right) S = (n_i + N_i)\left(\frac{\partial}{\partial t} + \frac{1}{E}\frac{\partial}{\partial x}\right) E + E\left(\frac{\partial}{\partial t} + \nu_i \frac{\partial}{\partial x}\right) N_i - En_i \frac{\partial \nu_i}{\partial x} + EY_i \tag{3.44}$$

Now let e be the electric field intensity in volts per meter induced across the biocell continuum. Also, we can state that

$$e = \frac{E}{x} \text{ across } D$$

implying electric field uniformity, where

D = biocell effective-width continuum
E = total electric potential across the cell continuum

From equation 3.44, we can write

$$\frac{T}{F}\left(\frac{\partial}{\partial t} - \frac{M}{D}\frac{\partial}{\partial x}\right) S = \frac{n_i + N_i}{D} + e^2\left(MD\frac{\partial N_i}{\partial x} - Mn_i\right) + D(n_i + N_i)\frac{\partial e}{\partial t} + eD\frac{\partial N_i}{\partial x} \tag{3.45}$$

or for $\partial e/\partial t = 0$, equation 3.45 becomes

$$e^2\left(MD\frac{\partial N_i}{\partial x} - Mn_i\right) + eD\frac{\partial N_i}{\partial x} + \left(\frac{n_i + N_i}{D} + \frac{TM}{FD}\frac{\partial S}{\partial x}\right) = 0 \tag{3.46}$$

Let

$$A = MD\frac{\partial N_i}{\partial x} \tag{3.47}$$

$$B = D\frac{\partial N_i}{\partial x} \tag{3.48}$$

$$C = \frac{n_i + N_i}{D} + \frac{TM}{FD}\left(\frac{\partial S}{\partial x}\right) \tag{3.49}$$

Hence, given the generalized solution of $e(x)$ in the steady state with space change and with $\partial e/\partial t = 0$, e from equation 3.46 is expressed as

$$e = \frac{1}{2A}[-B \pm \sqrt{B^2 - 4AC}] \tag{3.50}$$

or we can write

$$e = \frac{1/2}{MD(\partial N_i/\partial x) - Mn_i} - D\frac{\partial N_i}{\partial x} \pm \sqrt{\left(D\frac{\partial N_i}{\partial x}\right)^2 - 4\left(MD\frac{\partial N_i}{\partial x} - Mn_i\right)\left(\frac{n_i + N_i}{D} + \frac{TM}{FD}\frac{\partial S}{\partial x}\right)} \tag{3.51}$$

a special situation in which $\partial N_i/\partial x = \partial S/\partial x = 0$, implying that uniform N and Q will result:

$$e = \pm \frac{n_i + N_i}{Mn_i} \tag{3.52}$$

Therefore,

$$E = \pm \frac{D}{M}\left(1 + \frac{N_i}{n_i}\right) \text{ volts} \tag{3.53}$$

3.6 GENERAL SOLUTIONS OF *e*(*x*, *t*)[8]

Equation 3.51 could be rewritten as

$$a\frac{de}{dt} + be^2 + ce + d = 0 \tag{3.54}$$

where

$$a = D(n_i + N_i) \tag{3.55}$$

$$b = \left(MD\frac{\partial N_i}{\partial x} - Mn_i\right) \tag{3.56}$$

$$c = D\frac{\partial N_i}{\partial x} \tag{3.57}$$

$$d = \frac{n_i + N_i}{D} - \frac{T}{F}\left(\frac{\partial}{\partial t} - \frac{M}{D}\right)S \tag{3.58}$$

provided that $\partial S/\partial x$ is independent of t.

Therefore, equation 3.54 becomes

$$\frac{de}{be^2 + ce + d} = -adt \tag{3.59}$$

and

$$\int \frac{de}{(e + \alpha/\beta)/(e + \alpha'/\beta)} = at \tag{3.60}$$

or

$$-at = \frac{\beta}{\alpha - \alpha'}\ln\frac{\alpha' + \beta e(x, t)}{\alpha + \beta e(x, t)} + e(o, x) \tag{3.61}$$

$$\frac{\alpha}{\beta}, \frac{\alpha'}{\beta} = \frac{-D(\partial N_i/\partial x) \pm \sqrt{D^2(\partial N_i/\partial x)^2 - 4bd}}{2[MD(\partial N_i/\partial x - Mn_i]} \tag{3.62}$$

[8] Ibid.

$$b = MD\frac{\partial N_i}{\partial x} - Mn_i \tag{3.63}$$

$$d = \frac{n_i + N_i}{D} - \frac{T}{F}\left(\frac{\partial}{\partial t} - \frac{M}{D}\frac{\partial}{\partial x}\right) S \tag{3.64}$$

Solution of equation 3.60 for expressing the developed electric field inside the bioelectrochemical cell continuum could be a spectrum or a set of specific solutions depending upon the state parameter α/β, b, and d. Securing such solutions are left as practical problems to tackle in this chapter.

3.7 SUMMARY

The dynamic model for the biochemical fuel cell in the complex frequency domain was developed expressing the cell response of the terminal DC voltage with respect to the total energy content of the bioorganic fuel. The cell response was given in terms of the transform functions of the reactants concentration, entropy, internal energy, and equivalent bioelectrochemical reactions.

The dynamic model for the process of bioconversion of organic fuel in the complex frequency domain was secured, expressing the response of the total energy extracted by the developed hydrocarbon (C_mH_{2m+2}) with respect to the concentration of any effective reactant contained within the multicompound bioorganic fuel. The response was given in terms of the transform functions of entropy, enthalpy, implicit effect of concentration, and the direct influence of the logarithmic function of the developed hydrocarbon order, which is the reciprocal of activity property.

Establishing of dynamic model for the biochemical fuel cell was shown as being essential for securing required information for the steady-state as well as the transient operational performance of this power-generating device. This involved optimizing the power density, selection of the most promising kind of bioorganic fuel, type of electrolyte, and the rate of oxidization. The dynamic model developed a source for kinetic data regarding internal biochemical reactions, identifying sources of voltage drops due to polarizations and stability of efficiency of energy transformation.

Dynamic modeling for the process of bioconversion, to extract a potential and effective hydrocarbon (C_mH_{2+2m}) from a preliminary prepared biofuel, indicated the dominant role of the biofuel activity or its order (m) in stabilizing a desired level of energy content in the developed output hydrocarbon. The role of the hydrocarbon order was shown as an implicit logarithmic function.

The information presented in this chapter was purely theoretical and was based on mathematical and physical conceptions. Experimental work has to be performed to supplement the kinetic and dynamic characteristics that could be provided by the models established. However, the analytical base presented here could be considered as an introductory step for the eventual utilization of the biocell as a reliable alternative energy source carrying with it tremendous positive environmental benefits.

The transform function for the concentration of the ith reactant with respect to the mean velocity of the fluidic reaction medium was developed as were operational constraints regarding maximum hydrocarbon energy extraction from prepared refuse in time and space domain.

Solution for the electric field intensity across the biocell electrodes was obtained in explicit form, and is valid under any state of bioelectrochemical and thermodynamic coupling.

3.8 SOLVED EXAMPLES

1. In the presence of bioconversion of prepared refuse to synthetic fuel, if the medium of electrochemical reaction is incompressible fluid, revise the equation accordingly.

Solution From equation 3.43,

$$\frac{\partial n_i}{\partial t} + \overline{\nabla} \cdot n_i \overline{V}_i = \sum_{i=1}^{n} Y_i \tag{1.a}$$

Therefore,

$$\frac{\partial n_i}{\partial t} + n_i \overline{\nabla} \cdot \overline{V}_i + \overline{V}_i \cdot \overline{\nabla} n_i = \sum_{i=1}^{n} Y_i \tag{1.b}$$

For incompressible fluids,

$$\overline{\nabla} \cdot \overline{V}_i = 0 \tag{1.c}$$

Therefore, equation (1.b) becomes

$$\frac{\partial n_i}{\partial t} + \overline{V}_i \cdot \overline{\nabla} n_i = \sum_{i=1}^{N} Y_i \tag{1.d}$$

For a one-dimensional space change in n_i, equation (1.d) becomes

$$\frac{\partial n_i}{\partial t} + V_i \frac{\partial n_i}{\partial x} \partial x = \sum_{i=1}^{n} Y_i \tag{1.e}$$

or we can write,

$$\frac{\partial n_i}{\partial t} + V_i \frac{\partial n_i}{\partial x} = \sum_{i=1}^{n_i} Y_i \tag{1.f}$$

2. In the process of bioconversion of prepared refuse to synthetic fuel, the time development of the hydrocarbon of order m in terms of space change in the concentration of the ith reaction process is expressed by

$$\frac{\partial m}{\partial t} = \frac{1}{(RT_o)^2}\left[\frac{k_t - k_x}{\partial n_i/\partial t}\right] m$$

If k_t and k_x are constants and $\partial n_i/\partial x$ is linear with respect to x, solve for the time response of m.

Solution Given that

$$\frac{\partial m}{\partial t} = \frac{m}{(RT_o)^2}\frac{k_t - k_x}{\partial n_i/\partial x} \tag{2.a}$$

for

$$\partial n_i/\partial x = a_i + a_2 x. \tag{2.b}$$

Therefore,

$$\frac{\partial m}{m} = \frac{\partial t}{(RT_o)^2}\left(k_t - \frac{k_x}{a_1 + a_2 x}\right) \tag{2.c}$$

If m is a function of x and t, as the actual case is, equation (2.c) becomes

$$\ln m = \frac{t}{(RT_o)^2} - \left(k_t - \frac{k_x}{a_1 + a_2 x}\right) + m(x) \tag{2.d}$$

At $x = 0$, $m = M$

$$m(0) = \left[\ln m(0) - (k_t - \frac{k_x}{a_1}\right] \tag{2.e}$$

And at the other end of the process continuum, where $x = D$, $t = t_D$ and hence $m(x) = m(D)$, from equation (2.d)

$$\ln m(D) = \frac{t_D}{(RT_o)^2} + \left(k_t - \frac{k_x}{a_1 + a_2 D}\right) + m(D) \tag{2.f}$$

or

$$m(D) = \ln m(D) - \frac{t_o}{(RT_o)^2} + \left(k_t - \frac{k_x}{a_1 + a_2 D}\right) \tag{2.g}$$

Hence $m(x)$ in equation (2.d) could be secured from equations (2.e) and (2.g), leading to complete response from equation (2.d), where

$$m = \exp\left[\frac{t}{(RT_o)^2} + \left(k_t - \frac{k_x}{a_1 + a_2 x}\right) + m(x)\right] \tag{2.h}$$

where $m(x, t)$ is assumed a linear function with respect to x.

3. Simplify the solution of the voltage induced in the biocell $e(x, t)$ in equation 3.61 for $\alpha = 0$ and $e_o(o, x) = e_0$ (constant).

Solution From equation 3.61, rewritten below,

$$
\begin{aligned}
-at &= \frac{-\beta}{\alpha'} \ln \frac{\alpha' + \beta e(x, t)}{\beta e(x, t)} \\
&= \frac{-\beta}{\alpha'} \ln \left[\frac{\alpha'}{\beta e(x, t)} + 1 \right]
\end{aligned}
\tag{3.a}
$$

Therefore

$$1 + \frac{\alpha'}{\beta e(x, t)} = e \frac{\alpha' at}{\beta} \tag{3.b}$$

or

$$e(x, t) = \frac{\alpha'/\beta}{e(\alpha' at/\beta) - 1} \tag{3.c}$$

usually

$$e \frac{\alpha' at}{\beta} \gg 1 \tag{3.d}$$

Therefore,

$$e(x, t) = \frac{\alpha'}{\beta} e - \frac{\alpha' at}{\beta} \tag{3.e}$$

where α', β, and a expressed in equations 3.55 and 3.62 are functions of x.

4. Regarding the dynamic model of bioconversion expressed in equation 3.37 and Figure 3.2, solve for the time response of W (max. amount of energy release), ν_s, and n (under the following conditions: $H(S) = T_o S(s)$, $G(S) = 0$, and $n(s) = n_o/s$.

Solution Equation 3.37 now becomes

$$
\begin{aligned}
\frac{W(S)}{n(s)} &= SRT_o \ln \frac{M}{m} \\
&= SRT_o \ln A_i
\end{aligned}
\tag{4.a}
$$

Therefore,

$$W(S) = Rn_o T_o \ln A_i \tag{4.b}$$

and

$$W(t) = Rn_o T_o \ln A_i U_o(t) \tag{4.c}$$

where $U_o(t)$ is an impulse function, implying that a steady and constant level of reactant's concentration in the biocell continuum will lead to an impulsive surge in molar energy content.

3.9 REVIEW QUESTIONS

1. What are the basic energy principles implied in treating the processes of reactions in the bioelectrochemical cell?
2. State the mathematical condition for an incompressible fluidic continuum in the bioconversion process.
3. Modify equation 3.4 to account for the case of irreversibility with respect to the process of electric energy release. Comment on the status of N and E.
4. Reduce the equation on the total rate of concentration (eq. 3.43) to space variation in one dimension.
5. Express a relationship for the efficiency of the bioelectrochemical cell, taking into consideration the ideal state with respect to total enthalpy input.
6. Show the conclusion expressed by equations 3.29 and 3.30 were in effect $W_o = W_{PT}$.
7. Express the electron mobility in equation 3.47 in terms of developed electric field and other operational parameters of the bioelectrochemical cell.
8. Identify the parameters α and α' in equation 3.60. Comment on the validity of each regarding the state of reaction in the bioelectrochemical cell.
9. Explain the physical and mathematical differences among W_{ic}^o, W_{ic}, and W_v.
10. Correlate the activity of the ith reactant with respect to the general hydrocarbon compound C_mH_{2m+2}.
11. Comment on the value of the molecular activity of any reactant within the bioelectrochemical cell A_i at various places and times.
12. In the process of bioelectrochemical conversion of prepared synthetic fuel to electric energy, list all possible variables in the reaction process.

3.10 PROBLEMS

1. Considering the fact that the bioelectrochemical cell is an irreversible energy device, modify equation 3.8 in terms of the electric field developed inside the cell.
2. Refer to equation 3.43 expressing the status of the total rate of reactant's concentration, and solve for $n_i(x, t)$ under the conditions that $\Sigma_i Y_i$ is proportional linearly with respect to time and V_i is constant.
3. Equation 3.51 expresses the general solution for $e(x)$. Identify a specific solution for the following cases:

$$\frac{\partial N_i}{\partial x} = a_1 x^2 + b_1$$

$$\frac{\partial Q}{\partial X} = a_2 x - b_2$$

$$n_i = a_3 x - b_3$$

where a_1, a_2, a_3, b_1, b_2, and b_3 are constants. Discuss the implications of the solution for $e(x)$.

4. Obtain the general solution for $e(x, t)$ in equation 3.46 when the heat entropy release S is linearly increasing with respect to x and linearly decreasing with respect to (t). Discuss the electrochemical implications for a stable level for $e(x, t)$.
5. Considering the electrochemical state of reaction within the biocell as irreversible, modify equation 3.24 accordingly.
6. Modify equation 3.8 under these simultaneous conditions:
   a. To include the electrochemical potential of the ith reactant.
   b. dn/dt is constant.
   c. dn/dt is linearly decreasing.
   d. dQ/dt is constant.
   Solve for du/dt for zero initial conditions.
7. Given that

$$n(s) = \frac{a}{s}, \; N(S) = \frac{b}{s^2}, \qquad \text{and} \qquad Q(S) = \frac{k_1}{k_2 + S^2}$$

   obtain the specific function $G(S)$ and $G_1(S)$.
8. Using the data given in Problem 7, establish G' and G'' in the (S) domain.
9. From solutions for G, G_1, G', and G'' secured in Problems 7 and 8, establish the entire transform function for $E(S)/U(S)$.
10. From the transform of $E(S)/U(S)$ secured in Problem 10, establish the steady-state solution by using the final value theorem; namely,

$$\lim_{s \to 0} S \frac{E(S)}{U(S)} = \lim_{t \to \infty} f(t)$$

   $f(t)$ is the inverse Laplace transform of $E(S)/U(S)$.
11. Consider the bioelectrochemical cell continuum as obeying the perfect gas laws. Using the concepts stated in equations 3.25, 3.28 and 3.29 derive sequential relationships applicable to the concept of reactants' concentrations and reactants' lumped-energy contents.
12. Refer to the dynamic model for the bioconversion of prepared refuse to synthetic fuel shown in Fig. 3.2. If $H(S) = TQ(S)$ and $G(S) = A_i/S^2$, solve for the transient behavior of $W(S)/n(S)$ at zero initial conditions.
13. Figure 3.1 represents functional modeling of the biofuel cell in the complex frequency domain. Establish such model given the following:

$$G' = G'' = 1$$

$$U_o = k_1, \text{ a constant}$$

$$G_1 = \frac{k_2}{s^n}, \; k_s \text{ is a constant}$$

$$H_1 = \frac{k_3}{s^{n+1}}, \text{ a constant}$$

$$n = \text{a molar concentration}$$

Then find out the steady state response in the time domain for $e(t)$ if

$$U(s) = \frac{k_4}{s^2}$$

14. Figure 3.2 represents functional modeling for the process of bioconversion of prepared refuse to organic fuel.

Given:

$$\frac{C_1(S)}{D_1(S)} = 1$$

$$\frac{C_2(S)}{D_2(S)} = \frac{1}{s}$$

$$H(S) - TS(S) = \frac{1}{s^2}$$

$$n(S) = \frac{1}{s^3}$$

Secure the parametric solution for $W(t)$ for any constant initial conditions required in the process of bioconversion.

15. Establish the general solution for the voltage induced $e(x, t)$ in the electrochemical biocell from equations 3.46–3.49, where S and $\partial s/\partial x$ each is some function of time. Let $N_i(x)$ = constant. Assume zero initial conditions.

16. Repeat Problem 15 for $N_i(x) = A_iX - m_x$ and $\partial/\partial x = A_ie^{-xt} + S_{ox}$. S_o is a constant and $e(0, 0) = 0$.

17. From equation 3.51, extract $e(x, t)$, and then identify every possibility that such a solution may not exist with its electrochemical reasoning.

18. Given the continuum of the bioelectrochemical cell as incompressible and if space variations are in three dimensions, obtain the general solution for $e(x, y, z, t)$ when $\partial e/\partial t = 0$.

19. Repeat Problem 15 if space variations are in three dimensions.

20. Repeat problem 16 if

$$\frac{\partial Q}{\partial x} = A_i\bar{e}^{-xt} + S_{ox},\ N_i(x) = A_ix - m_x$$

$$\frac{\partial Q}{\partial y} = A_i\bar{e}^{-xt} + S_{oy},\ N_i(y) = A_iy - m_y$$

$$\frac{\partial Q}{\partial z} = A_i\bar{e}^{-xt} + S_{oz},\ N_i(z) = A_iz - m_z$$

and

$$e(0, 0) = 0$$

where S_{ox}, S_{oy}, and S_{oz} are the steady entropy level.

21. In the process of bioconversion of refuse to synthetic fuel followed by the release of electric energy upon oxidation, if data relevant to these parameters are given, namely, n_i, N_i, m, H, S, T, and P, such that each of them is independent of time but has some unspecified function of space in one dimension, with a zero or fixed initial condition required, establish an equivalent lumped-parameter circuit for the bioelectrochemical cell in the steady state.

22. Repeat Problem 21 if n_i, N_i, S, H, T, P, and m each is proportional to e^{-xt}.

3.11 REFERENCES

1. Angel, E., and R. Bellman. *Dynamic Programming and Partial Differential Equation,* edited by Richard Bellman. New York: Academic Press, 1972.
2. Colichman, E. L. "Preliminary Biochemical Fuel Cell Investigations." *IEEE Proceedings,* Vol. 51 (1963).
3. Crouch, D. A. and Werth, J. "Batteries and the Economics of Load Leveling in Large Power Systems." Paper presented at the Winter 1973 meeting of the Power Engineering Society of IEEE, New York.
4. Denno, K. "Bio-Electrochemical Conversion of Refuse to Energy." Paper presented at the 1979 Proceedings of the 25th Annual Technical Meeting of the Institute of Environmental Sciences, Seattle, Wash., April/May 1979, pp. 316–321.
5. Denno, K. "Modeling of Bio-Electrochemical Conversion of Refuse to Energy." *Proceedings of the 2nd International Conference on Mathematical Modeling,* St. Louis, Missouri, (July, 1979), Vol. 1, pp. 1031–1040.
6. Denno, K. "Simulating Criterion for Bio-Chemical Conversion of Refuse to Synthetic Fuel and Electric Power." *Proceedings of 3rd International Conference on Alternative Energy Sources,* Miami, December 1980.
7. Grubb, W. T., and L. W. Niedrach. "Fuel Cells." In *Direct Energy Conversion.* New York: McGraw-Hill Book Co., 1966.
8. Heath, C. E., and W. J. Sweeney. "Kinetics and Catalysis in Fuel Cells." In W. Mitchel, ed., *Fuel Cells.* New York: Academic Press, 1963.
9. Heath, C. E., and C. H. Worsham. "The Electrochemical Oxidization of Hydrocarbons in a Fuel Cell." In G. J. Young, ed., *Fuel Cells,* Vol. 2. Reinhold, 1963.
10. Kleinstreuer, C., V. Vasudevan, and T. Poweigha. "Mathematical Modelling and Computer Simulation of Methane Production from Anaerobic Digestion of Organic Material." *Proceedings of the 2nd International Conference on Mathematical Modelling,* St. Louis, Missouri (1979), pp. 1041–1051.
11. Lamb, Sir Horace. *Hydrodynamics,* 6th ed. New York: Dover Publications, 1932.
12. Liebhafsky, H. A. and Cairns, E. J. *Fuel Cells and Fuel Batteries.* New York: John Wiley & Sons, Inc. 1968.
13. Mantell, C. *Batteries and Energy Systems.* New York: McGraw-Hill Book Co., 1970.
14. Reikert, L. "The Conversion of Energy in Chemical Reactions." *Energy Conversion, an International Journal,* Vol. 15 (1975).
15. Schultz, E. B. "Natural Gas Fuel Cells for Power Generation." *Am-Gas Journal,* Vol. 188 (1961).

16. Sisler, F. D. "Electrical Energy from Biochemical Fuel Cells." *New Scientist,* Vol. 12 (1961).
17. Truitt, J. K. "Engineering Aspects of Molten Carbonate Electric Generators Using Hydrocarbon Fuels." In *Fuel Cells,* Chem. Eng. Tech. Manual, AIChE, 1963.
18. Woodson, H. W., and J. R. Melcher. *Electromechanical Dynamics,* Vols. I, II, and III. New York: John Wiley & Sons, 1968.

4

The Redox-Flow-Cell Power System (Feasibility and Magnetic Properties)

4.1 INTRODUCTION[1]

This chapter introduces feasibility aspects of the redox-flow-cell power system as a viable alternative energy source. The name redox is derived from the nature of the electrochemical reaction process, that the stages of reduction and oxidation occur in the chambers of the catholyte and anolyte during discharge and conversely throughout recharging process.

The presentation in this chapter commences by introducing the redox flow cell, discussing its role as a reliable, renewable energy source, noting the problems of performance to be resolved, and considering its size with respect to the conventional system of the same power capacity [13, 15]. The next discussion deals with data tabulation of electrochemical characterization and cost analysis of a redox flow cell established at NASA Lewis Research Center [13, 15]. The following section details the process of reduction and oxidation and their link to the quantity of the Gibbs free energy function and some aspect of quantum chemistry regarding magnetofluid interaction in the catholyte chamber. Some important relationships of magne-

[1] Reprinted, with permission, from NASA TECHNICAL MEMORANDUM: NASA TMX-71540, paper entitled: "Electrically Rechargeable Redox Flow Cell" by Lawrence H. Thaler, Lewis Research Center, Cleveland, Ohio
And

Reprinted, with permission, from NASA TECHNICAL MEMORANDUM: NASA TMX-3192, paper entitled: "Cost and Size Estimates For an Electrochemical Bulk Energy Storage Concept," by Marvin Warshay and Lyle O. Wright at Lewis Research Center, Cleveland, Ohio.

tization for paramagnetic and ferromagnetic solution relevant to the redox flow cell will be discussed together with the phenomenon of magnetic relaxation.

Then some detail about the process of magnetofluid interaction in the catholyte chamber of the redox flow cell is presented. Data from experimental work conducted at the Energy Research Lab at the New Jersey Institute of Technology (NJIT) will correlate the concepts of the gyromagnetic resistance, magnetization energy, the rate of magnetization, and relaxation. Also discussed is the meaning of magnetic susceptibility with respect to the concentration and velocity of the catholyte ferric solution in terms of the exciting current which is the source of the existing magnetic field. This concept of magnetic fluid interaction in the catholyte has been dealt with in some detail, mainly because an induced current developed during the process of discharge and recharge will act to reduce the redox flow-cell efficiency as well as its terminal voltage upon load increase connected to the cell. This theoretical and experimental analysis will serve as a guide for future work, geared for solving problems involving performance improvement of the redox flow cell and its compatibility with other modes of alternative energy sources [1, 2, 13, 15]. The redox flow cell operates on the principle of oxidation-reduction reaction in which the ions of the redox couple remain soluble in their electrolytes in either oxidized or reduced states.

The redox couple could be $|TiCl_3\text{-}TiCl_4|$ $|FeCl_3\text{-}FeCl_2|$ or $|Ti^{+3}\text{-}Ti^4|$ $|Fe^{+3}\text{-}Fe^{+2}|$.[2] These couples are usually pumped into the reaction chamber, which contains inert carbon electrodes separated by an anion-permeable selective ion exchange membrane. On discharge, the Fe^{+3} ions will be reduced to Fe^{+2}, and the Ti^{+3} is oxidated to Ti^4. The selective membrane could be a hydrogen ion (H^+) type of a chlorine (Cl^-) type. See Fig. 4.1.

The redox-flow-cell battery is electrically rechargeable by reversing the direction of flow of the electric current. Operating temperatures of the redox battery are usually at the ambient level not exceeding 80°C.

Studies conducted at the NASA Lewis Research Center indicated the vast promise of the redox flow system as very appealing to meet an acceptable application in the regime of peak power leveling.

Physical size of a redox power plant is on the order of 2 percent of a comparable pumped hydroelectric storage plant; the capital cost of a 10-MW, 60–85 MWh redox-flow system was estimated to range from $189 to $327 per kilowatt.[3]

Also the redox flow system possesses other potential benefits, including its low environmental impact, its high efficiency at low temperatures, its apparent absence of life-cycle limitations, and its fast response, which produces a power leveling system of vast potential [13, 15].

[2] Ibid.

[3] Ibid.

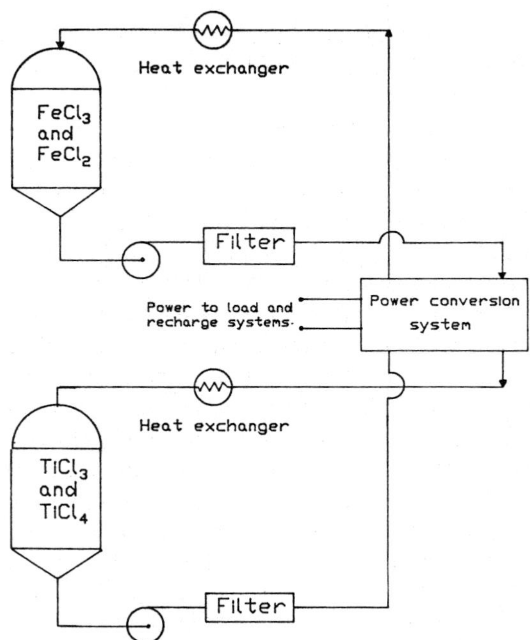

Figure 4.1 Electrically rechargeable redox-flow-cell power generation system. (Reprinted, with permission, from NASA TMY-71540 and NASA TMX-3192.)

However, several problems have to be resolved to achieve most efficient utilization of the electrically rechargeable redox flow system. Those problems are indeed interconnected and include optimization of the generating power density for the goal of 648 watts/m^2, reduction of the internal circulating current in the individual cells, and the optimum selection of the membrane.

In this chapter, the theoretical and experimental analysis presented deals with some of the causes of internal circulating currents that are in the catholyte and will be reduced later in the process of discharging to ferrous ionic fluid, resulting in the release of electrons and movement of H^+ from the anode compartment through the membrane to the cathode compartment, or negative ions Cl^- must transfer from the cathode chamber to the anode chamber.

Regardless of the type of membrane and solvent, the mere existence of iron fluid in the ionic state during the periods of charge and discharge points to the unique phenomenon of a dipolar flow of Fe^{+3} and/or Fe^{+2} subjected to the exposure of an applied magnetic field due to current flow to the external load during the period of discharge or due to current injection during charging process.

Interaction between the produced magnetic field with the dipolar fluid will generate magnetization losses and consequent internal currents induced in the iron fluid. This induced current is a consequence of Lenz's law, where the magnetic field resulting from the induced currents counteracts the applied magnetic field due to the discharge current that will be delivered to the load.

4.2 THE REDOX SYSTEM DESCRIPTION[4]

The redox flow cell consists of two compartments having separate different electrolytes and an inert carbon electrode separated by an anion-permeable ion exchange membrane. In the anolyte compartment, an aqueous electrolyte containing $TiCl_3$-$TiCl_4$ chemical solution will have to circulate from its storage tank into the anolyte and then back to its tank. In the catholyte compartment, an aqueous electrode containing $FeCl_3$-$FeCl_2$ chemical solution is to be circulated into the catholyte and back to its tank, as shown in Fig. 4.1.

During the process of discharge, where the cell is delivering electric power to its load, $FeCl_3$ is reduced to $FeCl_2$, while $TiCl_3$ is oxidized to $TiCl_4$. At the same time an electrochemical reaction is taking place involving passage of chlorine ions (Cl^-) between the anolyte and the catholyte through the exchange membrane. Also in some systems, hydrogen ion (H^+) membrane may be used instead of Cl^-.[5]

Fig. 4.2 shows detail of the electrochemical reactions with H^+ or Cl^- membrane.

Useful information [15] about electrochemical and size characteristics of the redox flow cell power plant is represented by data for a 10-MW, 85-MWh system of $|TiCl_3|$ $|TiCl_4|$ $|FeCl_3|$ $|FeCl_2|$ at an operating temperature of 80°C and 90 percent efficiency.

Examples of electrochemical data representative for a redox flow cell are: current densities ranging from 10.8 to 108 mA/cm^2 (for a total electrode area of 1.55×10^9 cm^2), voltage output per cell of 0.60 volts, and energy density of 13.25 watt-hour per kilogram weight of the cell.[6]

4.3 REDOX-FLOW-CELL NOMINAL VOLTAGE

Refer to Figure 4.2, based on the work performed at NASA Lewis Research Center by Lawrence Thaller, M. Warshay, and L. O. Wright [13, 15]. At the

[4] Ibid.

[5] Ibid.

[6] Ibid.

Cathode	$FeCl_3 + e^- \rightarrow FeCl_2 + Cl^-$
Anode	$TiCl_3 + Cl^- \rightarrow TiCl_4 + e^-$
Overall discharge reaction	$FeCl_3 + TiCl_3 \rightarrow FeCl_2 + TiCl_4$

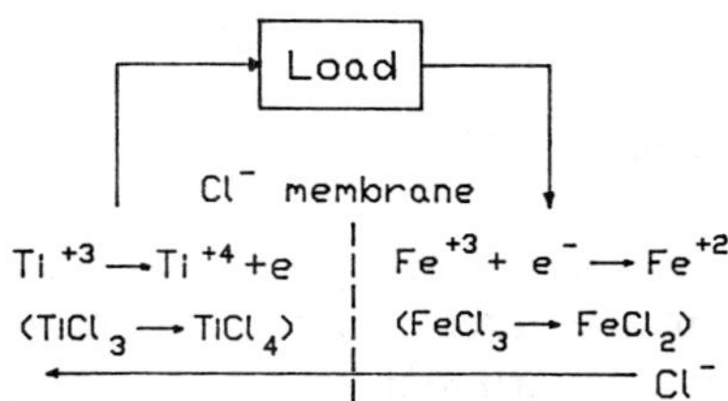

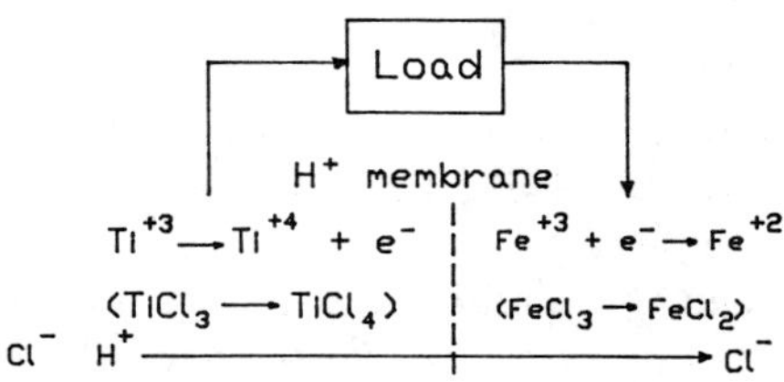

Figure 4.2 Electrochemical reactions for redox flow cell. (Reprinted, with permission, from NASA TMY-71540 and NASA TMX-3192.)

anolyte, the chemical reaction equation is

$$Ti^{+3} \rightarrow Ti^{4} + \bar{e} \text{ (a reduction reaction)} \tag{4.1}$$

while at the catholyte, the reaction equation is

$$Fe^{+3} + \bar{e} \rightarrow Fe^{+2} \text{ (an oxidation reaction)} \tag{4.2}$$

or the overall discharge reaction equation for the redox flow cell becomes

$$FeCl_3 + TiCl_3 \rightarrow FeCl_2 + TiCl_4 \tag{4.3}$$

If n moles of electrons are associated with the reaction, and the internal voltage produced is V_o, the electrical energy released per kilogram-mole is

$$W_e = nFV_o \tag{4.4}$$

However,

$$W_e \leq \Delta G \tag{4.5}$$

where

W_e = the electrical energy released per kilogram-mole in an orderly pattern
F = Faraday's constant = 96.5×10^6 coulombs
ΔG = the Gibbs free energy, defined as the maximum energy release at standard pressure and temperature

Or we can state also that

$$\Delta G = \sum_r \Delta G_r - \sum_p \Delta G_p \tag{4.6}$$

where

ΔG_r = the total change in Gibbs energy of all reactants
ΔG_p = the total change in Gibbs energy of all products

From equations 4.4 and 4.5,

$$V_o \leq \frac{\Delta G}{nF} \tag{4.7}$$

Now, we can calculate ΔG relevant to equation 4.7 [from data obtained from reference 15], where V_o is 0.6 volts.

From equation 4.2, $n = 1$; therefore,

$$0.6 = \frac{\Delta G}{1 \times 96.5 \times 10^6}$$

Hence,

$$\Delta G = 57.90 \times 10^6 \text{ joule/kg-mole}$$

at standard pressure and temperature.

4.4 ANALYTICAL THEORY OF MAGNETIZATION

The ferric dipolar medium in a liquid state resembles a continuum of freely rotating magnetic dipoles restrained by their total angular momentum quantum number which limits the degrees of freedom through which each dipole may align itself with the direction of an external magnetic field.

According to the classical theory of paramagnetism, the magnetic moment M per unit volume can be calculated using the same approach in accord with the Langvin-Debye theory,[7]

$$M = N\, L\left[\frac{\mu' H}{kT}\right] \tag{4.8}$$

[7] A. J. Dekker, *Solid State Physics*, © 1963, pp. 450, 454, 455. Reprinted by permission of Prentice-Hall, Inc., Englewood Cliffs, NJ.

where

$$L\left[\frac{\mu' H}{kT}\right] = \text{the Langevin function}$$

N = number of dipoles per cubic meter
μ' = magnetic dipole moment
H = external magnetic field
k = Boltzmann constant
T = absolute temperature in Kelvin

If $H \ll kT$, equation 4.8 becomes

$$M = \frac{N\mu'^2 H}{3kT}$$

However by drifting slightly into the theory of quantum mechanics, paramagnetism limits the degrees of freedom that a dipole may align itself with respect to an external field, depending on the value of its total angular momentum quantum number J.

We can recall from quantum chemistry that the quantum number (J) combines the total orbital angular momentum (L) and the total spin (S) of the ion electronic system.

With respect to statistical mechanics, the magnetization M can be expressed as[8]

$$M = N \frac{\sum_{-J}^{+J} M_J g \mu_{\text{eff}} \exp(M_J g \mu_{\text{eff}} H/kT)}{\sum_{-J}^{+J} \exp(M_J g \mu_{\text{eff}} H/kT)} \tag{4.10}$$

where

$M_J = J\ (J - 1), \ldots -(J - 1), -J$ = the magnetic quantum number associated with J (4.11)
μ_{eff} = effective total magnetic moment for the ionic dipole associated with any value of J
$\mu_{\text{eff}} = g\mu_B\sqrt{J(J + 1)}$ (4.12)

where

J = total angular momentum quantum number
g = gyromagnetic factor which is the ratio of magnetic moment to the angular momentum of the dipole
μ_B = Bohr magnetron = $-eh/2mc = 0.927 \times 10^{-20}$ erg/oersted
= 11.75×10^{-30} joules/ampere-turns/meter
e, m = the electronic charge and mass, respectively
h = Planck constant
c = the velocity of light

[8] Ibid.

Also,[9]

$$g = 1 + \frac{J(J+1) + S(S+1) - L(L+1)}{2J(J+1)} \tag{4.13}$$

Then in accord with Hund's rules,

1. In an incompletely filled shell,

   $J = (L - S)$ for a shell less than half occupied

   $J = (L + S)$ for a shell more than half occupied

2. The electron spins add to give the maximum possible (S), consistent with Pauli's principle.
3. The total orbital momenta combine to give the maximum value for L consistent with Pauli's principle.

It has been verified that $\mu_{\text{eff}}H \ll kT$ is valid for the present ferric solution at room temperature, with a calculated value of $J = 5$ for Fe^{+3} ions. Therefore, the solution M, expressed in equation 4.10, could become as in equation 4.9 with μ replaced by $\mu_{\text{eff}} = g\mu B\sqrt{J(J+L)}$.

4.5 MAGNETIZATION AND RELAXATION OF FERROMAGNETIC SOLUTION[10]

This phenomenon is expressed by [2]

$$\frac{dM}{dt} + \frac{2KT}{\nu}M = \frac{2\pi N M_s^2 V_h^2}{\nu} H(t) \int_{-1}^{+1} (1 - x^2) F(x) dx \tag{4.14}$$

and

$$\tau = \frac{\nu}{2kT} \text{ as the time constant of magnetization}$$

where

M = magnetization function
$F(x)$ = magnetites distribution function in terms of space variable x
ν = rotational mobility
$\nu = 6\rho V_h$
ρ = fluid viscosity
N = concentration of magnetite particles per unit volume

[9] Ibid.

[10] Journal of Applied Physics, Vol. 49, No. 6, pp. 3422–3429. Paper entitled: "Dynamic Magnetization in Ferrofluids" by H. Bogardus, D. A. Kruger and D. Thompson. Published, June 1978 by American Institute of Physics.

V_h = hydrodynamic volume = $\frac{1}{6}\pi D^3$
D = magnetite diameter
k = Boltzmann constant
M_s = saturation magnetization
T = absolute temperature, and a function of time
$H(t)$ = time varying applied magnetic field function

For ferric chloride solution, the viscosity is on the order of $0.07p$, and the corresponding typical value for the hydrodynamic diameter is 2×10^{-8} cm.[11] Rotational mobility of a magnetite particle in ferric chloride solution is

$$(0.07)(2 \times 10^{-8})^3 = 1.76 \times 10^{-24}$$

At $T = 300$ K,

$$\tau = \frac{\nu}{2kT} = 20 \times 10^{-6} \text{ sec}$$

Now let

$$R = \frac{2\pi n M_s V_h^2}{\nu} \int_{-1}^{+1} (1 - x^2)F(x)dx \tag{4.16}$$

Equation 4.14 becomes

$$\frac{dM}{dt} + \frac{M}{\tau} = RH(t) \tag{4.17}$$

The magnetization response $M(t)$ due to a rectangular pulse $H(t)$ is

$$H(t) = H_o[U_{-1}(t) - U_{-1}(t - Tp)] \tag{4.18}$$

with the initial condition that $M(o) = 0$.[12]

Equation 4.17 becomes

$$\frac{dM}{dt} + \frac{1}{\tau} M(t) = RH_o[U_{-1}(t) - U_{-1}(t - Tp)] \tag{4.19}$$

Taking the Laplace transformation of equation 4.19 followed by the inverse transform, yields the response solution for $M(t)$:

$$M(t) = RH_o[U_{-1}(1 - e^{-t/\tau}) - U_{-1}(t - Tp)(1 - e^{t-Tp/\tau})] \tag{4.20}$$

[11] Ibid.

[12] © 1985 Hemisphere Publishing Corp. Reprinted, with permission, from Alternative Energy Sources VI, Vol. 3, pp. 321–24, 1985. Paper entitled: "More on the Parametric Transformation of Water-Based Ferromagnetic Solutions for Energy Conversion" by A. Velingker, K. Sohn & K. Denno. Edited by T. Nejar Veziroglu.

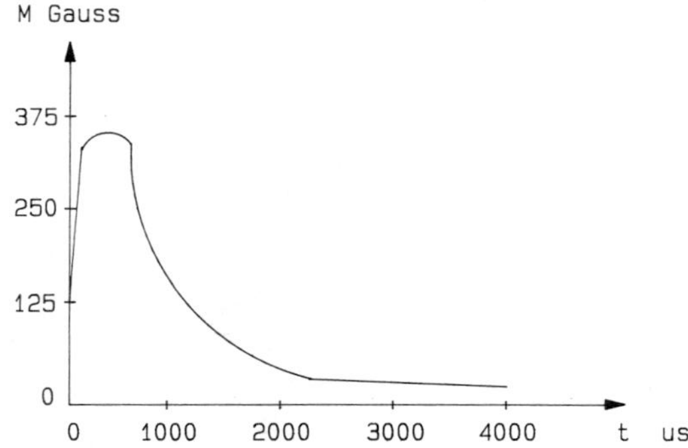

Figure 4.3 Magnetization for 16% $FeCl_3$ solution at saturation level. $pT = 500\ \mu sec$, $H = 150$ Oe. ("More on Parametric Transformation of Water-Based Ferromagnetic Solutions for Energy Conversion" by A. Velingker, K. Sohn and K. Denno, published in *Alternative Energy Resources* by Hemisphere Publishing Company (1985) New York.)

where τ_p is the time width of the field pulse. The plot of $M(t)$ expressed by equation 4.20 is shown in Figure 4.3.

Direct linkage of the time constant τ in equation 4.14 points out the role of temperature in the evolution of the magnetization function $M(t)$ with respect to temperature T. Equation 4.19 can be rewritten in a form compatible to that of equation 4.14,

$$\frac{dM}{dt} + \frac{2kT(t)}{\nu} = RH(t)^{13} \tag{4.21}$$

In equation 4.21 the temperature T is expressed as a function of time t, since the fluid temperature will rise with time due to the gyromagnetic energy expended upon the application of $H(t)$.

Solution of equation 4.21 by the quadrature formula procedure gives the following solution for M:

$$M = c \int_e \frac{-2kT(t)}{\nu}\, dt - \int_{e^-} \frac{2KT(t)}{\nu}\, dt = RH(t) \int_e \frac{2KT(t)}{\nu}\, dt \tag{4.22}$$

where c is a constant to be determined upon the initial state of magnetization. It can be noted from equation 4.22 that the transient response of M will not be affected, as it will take some time for the temperature of the ferrofluid to change the temperature T in a duration that is much larger than the typical value of τ.

[13] Ibid.

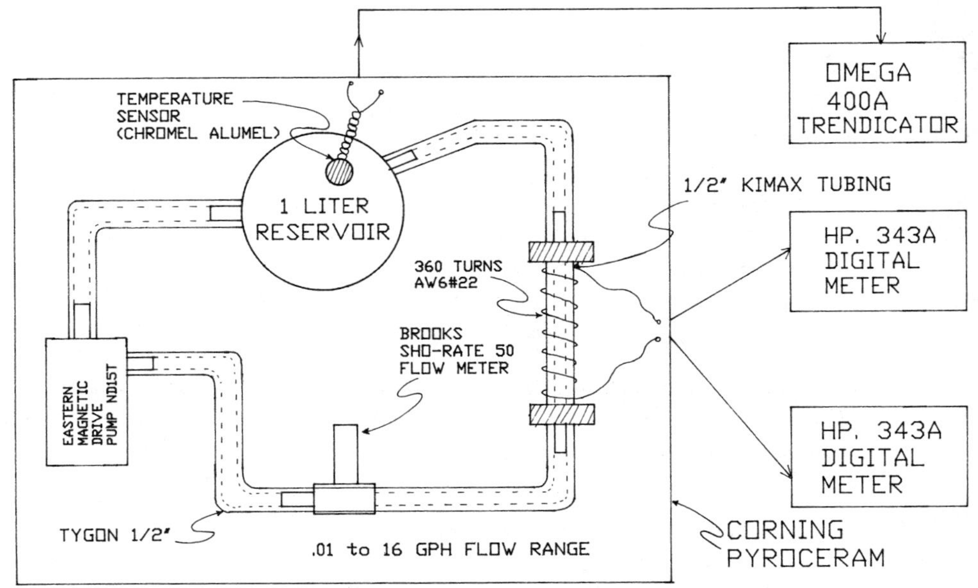

Figure 4.4 Experimental flow apparatus for ferric solution–magnetic field interaction. (© 1981 IEEE Reprinted with permission from 15th Electrical/Electronics Insulation Conference, October 19–22, 1981, Rosemont, Ill., pp. 226–230.)

4.6 CALCULATION OF QUASI-STATIC ENERGY DENSITY ACTING ON Fe^{+3} Dipolar Fluid[14]

As illustrated in Figure 4.4, the flow of Fe^{+3} dipolar liquid is through a cylindrical channel of a nonmagnetic wall surrounded by a 360-turns solenoid.

Upon the passage of DC current through the solenoid, an axial magnetic field will act on each dipole moving through the cylindrical channel. Those Fe^{+3} dipoles are floating within the solution and are rotating freely before their entrance into the magnetic field region.

However, we have to state that due to the complex nature of the ion clusters in their combination with chlorine ions, a different ion distribution may exist in the solution with respect to mass, mobility, and their probability of alignment with the external magnetic field. It can be said that since each ion has a total angular momentum quantum number of $J = 5$, each ferric ion can have five possible directions of orientation or degrees of freedom with the influence of the outside external magnetic field.[15]

[14] K. Denno, "Problems in the Redox Flow Cell Power System for Bulk Energy Storage," from Proceedings of 1979 Annual Technical Meeting, Institute of Environmental Sciences, pp. 226–230.

[15] Ibid.

Modes of motion (laminar or turbulent) can disrupt the process of magnetization with varying degrees.

Exchange coupling among dipoles, although weak in this fluidic solution, may impose some marked effect on the once already partially or even completely aligned dipoles.

Hence because of the ambiguous situation that seems to exist during the process of magnetization, the calculation will be confined only to provide information about the static energy density in joules per unit volume to which the dipoles will find themselves subjected, once they are exposed to the external magnetic field. Then it occurred as anticipated that the experimental work provided information relevant to the relative macroscopic movement of the magnetic dipoles by correlating the energy density obtained theoretically and the measured gyromagnetic power obtained experimentally.

Indeed equation 4.10 is the exact form to obtain information for magnetization of the dipolar fluid. But since $\mu_H \ll kT$ and the fact that all dipoles rotate freely in the solution, equation 4.9 can be considered very fairly to represent the magnetization imposed by an outside field H. Therefore, for the quasi-static energy density E_i,[16]

$$E_i = -\overline{H} \cdot \overline{M} \tag{4.23}$$

$$= -\frac{N\mu_{\text{eff}}^2 H^2}{3kT} \text{ joules/m}^3 \tag{4.24}$$

and H is the external magnetic field in ampere-turns/meter.

In the present simulated experimental setup, we can recall that the field H of a solenoid carrying current i is

$$H = \frac{n_i}{\rho} \text{ ampere-turns/meter} \tag{4.25}$$

where

n = number of turns of the solenoid
ρ = axial length of the solenoid

Therefore,

i = electric current in amp.

$$E = -\frac{N\mu_{\text{eff}}^2 (n^2 i^2)}{3kT^2} \text{ joules/m}^3 \tag{4.26}$$

N = concentration of ionic dipoles per cubic meter
i_L = current flowing through the solenoid producing and external axial magnetic field

[16] Ibid.

In the redox flow cell, i_L is the actual load current that is to be delivered to the load.

4.7 EXPERIMENTAL CATHOLYTE SIMULATION OF MAGNETOIONIC INTERACTION[17]

Magnetofluid interaction apparatus was set up in the Energy Conversion Laboratory at NJIT as shown in Figure 4.4. Ferric solution channel flow of glass material enclosed within a 360-turns solenoid was used as the medium of interaction.

Through the application of several distinct values of electric current, generating the corresponding magnetic field (H = Ni/l) with respect to one level of concentration at a time, and for flow velocities of 1 cm/sec and 2 cm/sec, data were secured for the gyromagnetic resistance at concentration levels ranging from 6 to 10 percent. These data are shown in Figures 4.5a through 4.5e for the two velocity levels indicated. The corresponding behavior of the expended quantum energy during the time of travel of ferric solution through the solenoid channel was obtained and plotted in Figures 4.6 and 4.7 (i.e., gyromagnetic power).

Furthermore, to present an account for distinct boundaries between magnetite alignment and magnetic relaxation, data have been compiled to establish a concentrational level at which saturation reversal takes place at flow velocities of 1 cm/sec and 2 cm/sec, respectively, as shown in Figures 4.7a and b.

And, finally, distribution patterns of the peak gyromagnetic resistance with respect to variable concentration obtained at 1 cm/sec and 2 cm/sec flow velocity, respectively, are shown in Figures 4.8a and 4.8b.

Let us look at what can be deduced from this work. The experimental data and the results demonstrated in the plots of Figures 4.5–4.8 verify the fact that the quantum nature characterizing the magnetofluid interaction of the water-based ferric solution at ambient temperatures with the exciting local stationary magnetic field is valid.

The following observations can be drawn:

1. The pattern of change of quantum gyromagnetic resistance, R_{gm}, increases nonlinearly with respect to the applied magnetic field, thus attaining open-circuit level at a maximum of about 2 amp with gradual reduction in current as concentration increases (see Figure 5e).

[17] © 1981 IEEE. Reprinted with permission, from 15th. ELECTRICAL/ELECTRONICS INSULATION CONFERENCE, October 19–22, 1981, Rosemont, IL. pp. 226–230. Paper entitled: "Spatial Distribution of Quantum Parameters in the Continuum of Ferromagnetic Liquid Insulation" by K. Denno.
(Source for Figs. 4.4–4.8b).

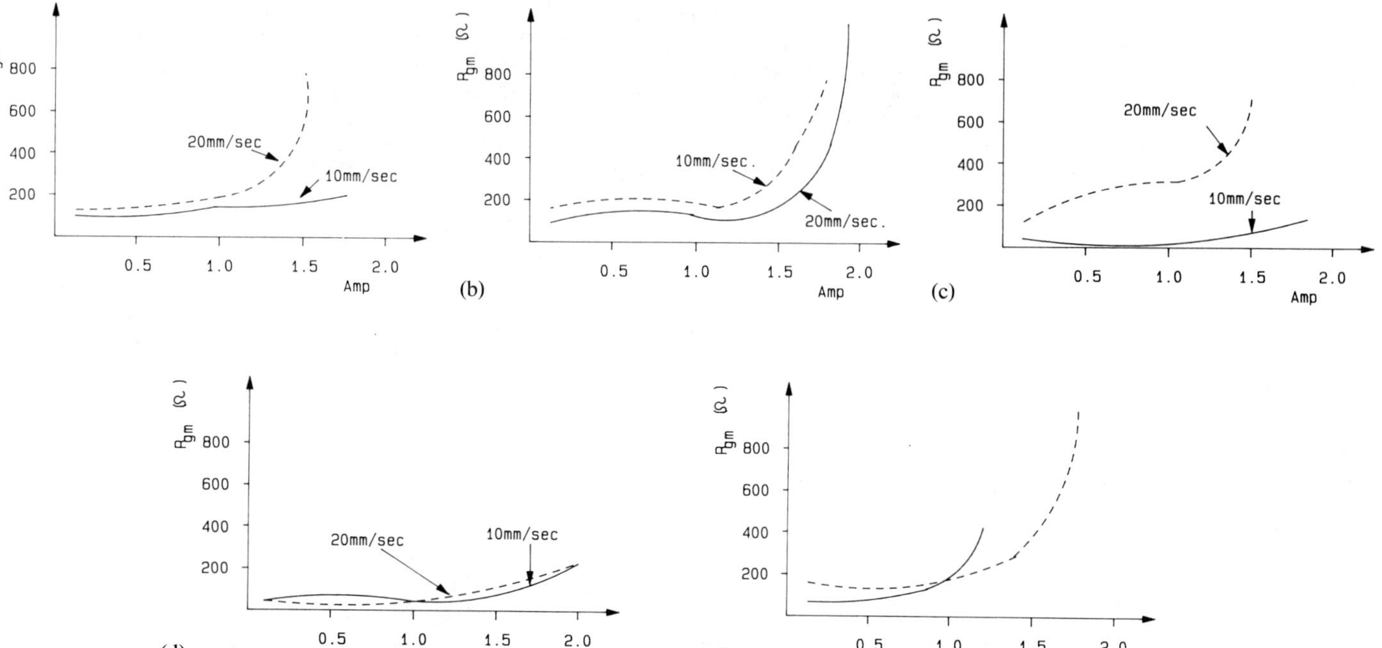

Figure 4.5.a Gyromagnetic resistance versus field current for 6% ferric chloride. (© 1981 IEEE. Figure parts reprinted with permission from 15th Electrical/Electronics Insulation Conference, October 19–22, 1981, Rosemont, Ill., pp. 226–230.)

Figure 4.5.b Gyromagnetic resistance versus field current for 7% ferric chloride.

Figure 4.5.c Gyromagnetic resistance versus field current for 8% ferric chloride.

Figure 4.5.d Gyromagnetic resistance versus field current for 9% ferric chloride.

Figure 4.5.e Gyromagnetic resistance versus field current for 10% ferric chloride.

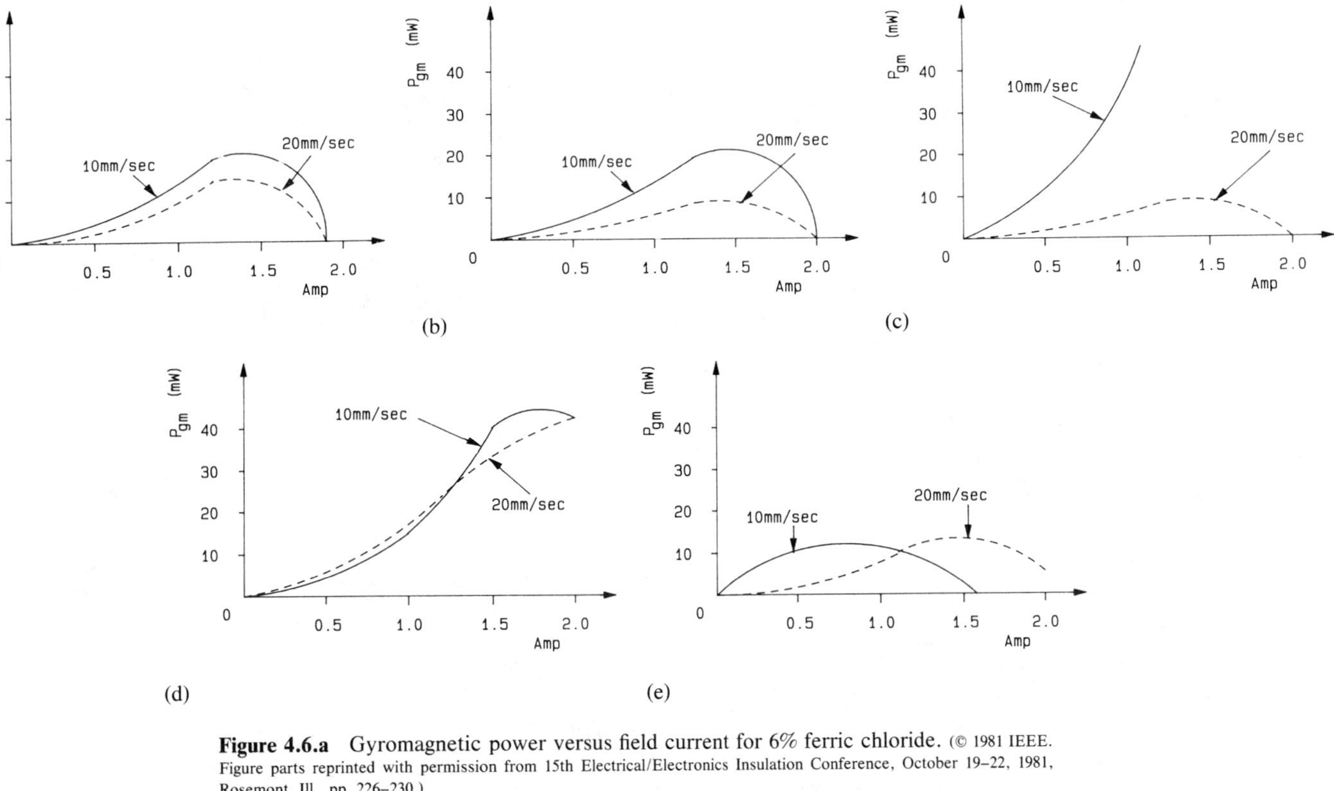

Figure 4.6.a Gyromagnetic power versus field current for 6% ferric chloride. (© 1981 IEEE. Figure parts reprinted with permission from 15th Electrical/Electronics Insulation Conference, October 19–22, 1981, Rosemont, Ill., pp. 226–230.)

Figure 4.6.b Gyromagnetic power versus field current for 7% ferric chloride.

Figure 4.6c Gyromagnetic power versus field current for 8% ferric chloride.

Figure 4.6.d Gyromagnetic power versus field current for 9% ferric chloride.

Figure 4.6.e Gyromagnetic power versus field current for 10% ferric chloride.

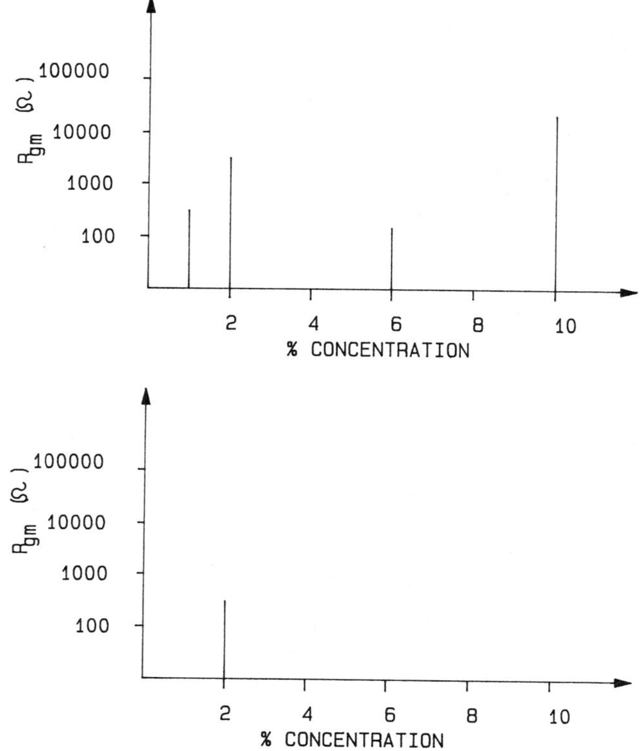

Figure 4.7.a R_{gm} inversion versus concentration for fluid velocity, 10 mm/sec. (© 1981 IEEE. Figure parts reprinted with permission from 15th Electrical/Electronics Insulation Conference, October 19–22, 1981, Rosemont, Ill., pp. 226–230.)

Figure 4.7.b R_{gm} inversion versus concentration for fluid velocity, 20 mm/sec.

2. With respect to gyromagnetic energy absorbed throughout the interaction, ferroresonance occurrence is confirmed at a decreasing level of applied magnetic field as concentration increases.

3. The regime of ferric concentration for sudden transfer from magnetization into relaxation occurs at 2, 6, and 10 percent, for flow velocity of 1 cm/sec, although it occurs at only 2 percent for a flow velocity of 2 cm/sec. These regimes are shown in Figures 4.8a and b.

4. Distributions of maximum quantum resistance versus concentrations shown in Figures 4.7.a and b resemble closely the spatial particles' distribution function with the attributed phenomena of magnetic alignment or relaxation to the role of magnetite rotation and disorientation, which are coupled with a slow dominant rotational time constant of the order of 1 mm/sec.

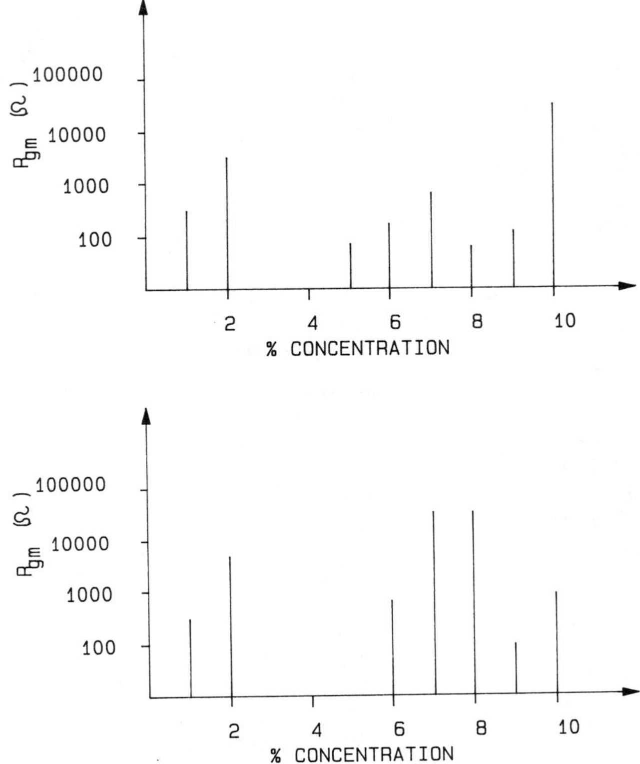

Figure 4.8.a R_{gm} inversion versus % concentration for fluid velocity, 10 mm/sec. (© 1981 IEEE. Figure parts reprinted with permission from 15th Electrical/Electronics Insulation Conference, October 19–22, 1981, Rosemont, Ill., pp. 226–230.)
Figure 4.8.b R_{gm} inversion versus % concentration for fluid velocity, 20 mm/sec.

The new central element in these results is the role of the variable concentration of ferric material within a water-based solution at ambient temperature. Varying ferric solution concentrations are compatible with the pattern of the spatial distribution function for magnetites in response to the applied magnetic field.

5. The persistent mode of ferroresonance expressed by the pattern of quantum energy with respect to exciting field, *identifies* the threshold limit at which the magnetic dipole moment per unit volume changes from magnetization regime into relaxation and concurrently establishes the doublet functional form of the magnetic susceptibility and eventually magnetic permeability.

Results reflecting ferroresonance indicate consistent occurrence of complete magnetization or relaxation where peak quantum energy has been expended and associated with open-circuit value of gyromagnetic resistance at almost the same magnetic field applied.

4.8 PARAMAGNETISM AND FERROMAGNETISM PHENOMENA IN THE CATHOLYTE[18]

Paramagnetism could be expressed by these basic equations:

1. Classical concept

$$M = N\mu' L\left(\frac{\mu H}{kT}\right) \tag{4.27}$$

or for $\mu' H \ll kT$

$$M = \frac{N\mu'^2 H}{3kT} \tag{4.28}$$

where

M = magnetization of dipole moment per unit volume
H = applied magnetic field
μ' = freely rotating dipole moment
N = density of dipoles per unit volume
k = Boltzmann constant
T = absolute temperature in Kelvins
L = Langevin-Debye function

$$\overline{M} = \chi \overline{H} \tag{4.29}$$

W_M = energy need for magnetic alignment or gyromagnetic energy expended

$$W_M = -\overline{\mu}' \cdot \overline{H} \tag{4.30}$$

$$\mu = 1 + 4\pi\chi \tag{4.31}$$

χ, μ = magnetic susceptibility and permeability, respectively

2. According to quantum approach,

$$M = NgJ\mu_B B_J(Z) \tag{4.32}$$

where

$$Z = \frac{gJ\mu_B H}{kT} \tag{4.33}$$

[18] Dekker, SOLID STATE PHYSICS, Chapter 18.

and

g = gyromagnetic factor
J = total angular momentum quantum number
μ_B = Bohr magneton
B_J = Brilloun function, which is expressed below in terms of J and Z:

$$B_J(Z) = \left[\frac{1 + 2J}{2J}\right] \coth\left[\frac{2J + 1}{2J}\right] Z - \frac{1}{2J} \coth\left[\frac{Z}{2J}\right]$$

Therefore,

$$\frac{dM}{dz} = N_g JB\left[-\left(\frac{1 + 2J}{2J}\right) \operatorname{csch}^2\left(\frac{1 + 2J}{J}\right) Z + \left(\frac{1}{2J}\right) \operatorname{csch}^2 \frac{Z}{2J}\right) \tag{4.34}$$

Ferromagnetism is characterized by the existence of domains having spontaneous magnetization due to an internal molecular field and also a net residual magnetism for the whole continuum.[19] Ferromagnetism is guided by the following basic relationships, which are based on quantum concepts:

$$\overline{H}_m = \overline{H}_a + \gamma M \tag{4.35}$$

Rewriting $\overline{M}$,

$$M = NgJ\mu_B B_J(Z) \tag{4.36}$$

Hence

$$\chi = \frac{M}{H_a} \tag{4.37}$$

or

$$\chi = g\mu_B(\overline{H}_a + \overline{M}) \frac{J}{kT} \tag{4.38}$$

where

$\overline{H}_m$ = molecular field
$\overline{H}_a$ = applied field
γ = Weiss constant

As mentioned earlier, ferromagnetic materials are characterized by the existence of residual or remanent magnetization, which is the amount of magnetic induction B for zero magnetic intensity H as demonstrated in the well-known hysteresis loop.

[19] Ibid.

4.9 TRANSPORT PROPERTIES OF CATHOLYTE FERRIC SOLUTION[20]

The foregoing equations for ferromagnetism apply 100 percent for solid material where the magnetic dipoles experience great rigidity in contrast to ferromagnetic liquids. In ferromagnetic liquids the dipoles are very loose and can be assumed to be freely rotating (similar to paramagnetic materials). Therefore, with respect to determining the pattern of change for the dipole moment per unit volume and the energy spent through magnetization, the fact of loose dipoles has been used.

From previous work carried out by this author, the mode of distribution for the magnetizing energy absorbed by the freely rotating magnetic dipoles versus exciting load current has been established at various levels of fluid velocity ranges. The energy mode of distribution showed clearly the phenomenon of resonance around a close, unified value of the exciting source current. The experimental setup has been shown in Figure 4.4. The ferric solution was adjusted to flow at various linear velocities through a cylindrical channel enclosed by a current-carrying solenoid. The channel is of 18-cm axial length and 12.5-mm diameter with the surrounding solenoid of 360 turns.

For each fluid velocity with the circuit energized by a stable DC power supply, readings were taken by accurate digital meters for the net voltage drop across the solenoid terminals, total circuit current, and a precise measurement for the solenoid resistance by a digital ohm-meter at the same moment that other measurements were taken. From those data the gyromagnetic power supplied from the source P_{gm} is obtained from the equation[21]

$$P_{gm} = P_{in} - \frac{V_L^2}{R_L} \tag{4.39}$$

where

P_{in} = total power input to the circuit
V_L, R_L = terminal voltage and DC resistance of the solenoid, respectively

The gyromagnetic current i_{gm} is obtained from the following equation,

$$P_{gm} = i_{gm}^2 R_{gm} \tag{4.40}$$

R_{gm} = the gyromagnetic resistance of the ferric flow

Further work has been carried out to identify the state of magnetization or the dipole moment/unit volume with respect to the applied magnetic field:

[20] K. Denno, "Problems in the Redox Flow Cell Power System for Bulk Energy Storage," presented at the 1979 Annual Technical Meeting, Institute of Environmental Sciences and published in the Proceedings. Reprinted with permission.

[21] Ibid.

$$M = \frac{dW_{gm}}{dH_a} \tag{4.41}$$

where W_{gm} and H_a are the calculated gyromagnetic energy and applied magnetic field, respectively.

Also from the pattern of change for the state of magnetization, the magnetic susceptibility has been calculated with respect to the external magnetic field. where

$$\chi = \frac{dM}{dH_a} \tag{4.42}$$

Curves representing the gyromagnetic energy W_{gm}, the magnetic dipole moment per unit volume m, and the magnetic susceptibility χ versus H_a are plotted in Figure 4.9a through e, with respect to a range of linear flow velocities as detailed in the paragraphs that follow.

Now from equations 4.27–4.30, the applied energy density for the ferromagnetic fluid could be expressed as

$$W_{gm} = \frac{N\mu'^2H^2}{3kT} \tag{4.43}$$

and since the field H is generated by a solenoid of n turns, length l carrying current i (as indicated in [6]), with $H = ni/l$, from equation 4.43

$$W_{gm} = \frac{N\mu^2i^2n^2}{3kTl^2} \tag{4.44}$$

Then if P_{gm} is the gyromagnetic power that can be measured experimentally [11] and W_{gm} can be calculated from equation 4.44,

$$W_{gm} = P_{gm}\frac{dM}{dt} \tag{4.45}$$

where dM/dt is the linear volumetric magnetic displacement per second.

Equation 4.45 can also be written in another form:

$$W_{gm} = (P_{gm})(V)(A) \tag{4.46}$$

where V and A are the linear velocity and effective cross section of the ferromagnetic fluid, respectively.

4.10 MAGNETIC RESONANCE AND SUSCEPTIBILITY[22]

The analysis presented up to now centers on the magnetic treatment of the dipolar ferromagnetic liquid, which is characterized by spontaneous magnetization for a domain and residual field for the total continuum. The fact is

[22] K. Denno. "Magnetic Transport Properties of Dipolar Conducting Liquids at Ambient Temperature," Journal of Electrostatics No. 7, 1979 pp. 345–360.

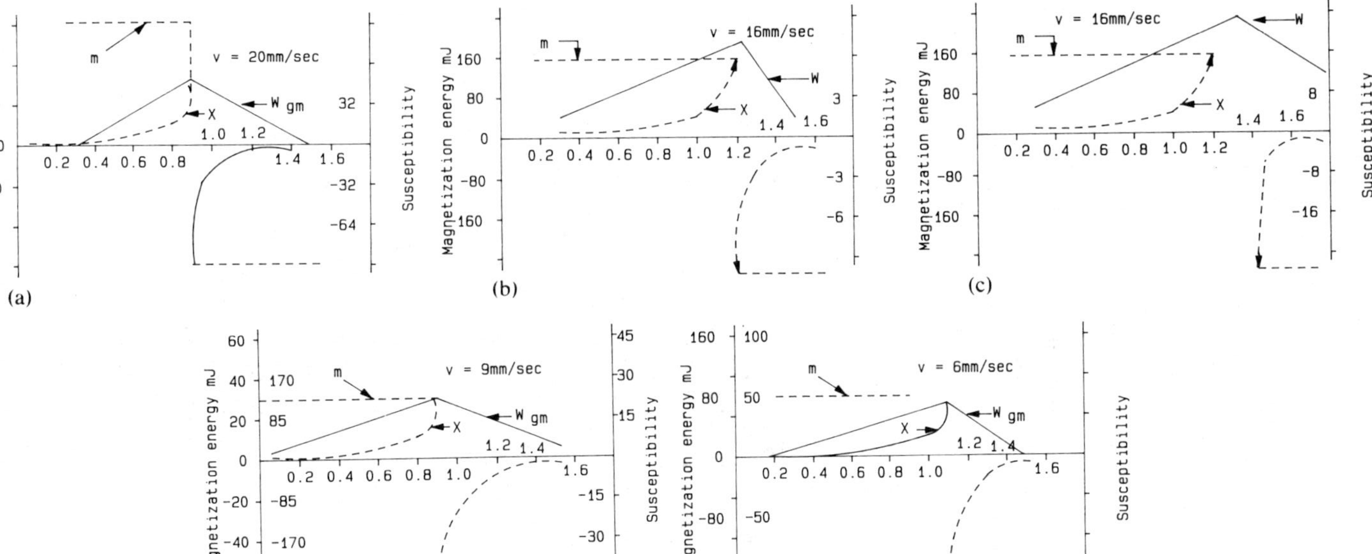

Figure 4.9.a Magnetization energy, magnetic dipole moment per centimeter, and susceptibility versus H. (All figure parts from "Magnetic Transport Properties of Dipolar Conducting Liquids at Ambient Temperature" by K. Denno for *Journal of Electrostatics*, 7, 1979. © Elsevier Scientific Publishing Company, Amsterdam, 1979.)

Figure 4.9.b Magnetization energy, magnetic dipole moment per centimeter, and susceptibility versus H.

Figure 4.9.c Magnetization energy, magnetic dipole moment per centimeter, and susceptibility versus H.

Figure 4.9.d Magnetization energy, magnetic dipole moment per centimeter, and susceptibility versus H.

Figure 4.9.e Magnetization energy, magnetic dipole moment per centimeter, and susceptibility versus H.

that various domains in the continuum are bounded by Bloch walls that can be displaced by the application of an external magnetic field. Theoretical treatment for the transport properties of ferromagnetics, whether in the solid or liquid state, has to follow the quantum concept, especially with respect to the fact that the total magnetic field acting on any dipole must come from the combination of an external component and the internal cooperative effect contributed by the state of magnetization associated with the Weiss constant.

By now, we may come to grasp the physical reality that the liquid dipoles are freely rotating and resemble a similar situation that exists with paramagnetic materials. The ferromagnetic fluid is at an ambient temperature, so that the Langevin function can be approximated as a classical function; that is,

$$L(\mu' H/kT) \approx \mu'^2 H/3kT$$

Another fact that justified treating the dipolar properties of the ferric solution as being close to a paramagnetic liquid is that experimental work indicated clearly a very loose presence of Curie temperature.

On the other hand, a quantum aspect of this ferromagnetic liquid is still adhered to, namely, its limited degrees of freedom with respect to directions of magnetic orientations that are set by the total angular momentum quantum number J. Therefore, it is indicated that studying transport properties of ferromagnetic liquid focuses on a combined theoretical concept linking the classical paramagnetic theory with the quantum basis of ferromagnetics.

Now coming to the results shown in Figures 4.5 and 4.6, resonance is a demonstrated fact for the gyromagnetic energy versus the external applied magnetic field. Resonance resembling the maximum expended energy is shown to occur around a field value of 5.55×10^3 At/cm for most cases of fluid linear velocities.

This value for the external magnetic field corresponds to the level of saturation. Slight deviations of saturation level from the specific value of 5.5×10^3 At/cm are attributed to the intervening effects of preceding magnetization as well as remagnetization and relaxation.

The next dipolar property to be demonstrated is the state of magnetic dipole moment per unit volume m, which shows a sudden inversion from a positive DC value to a symmetrical negative level with respect to the applied magnetic field. This discontinuous change occurs at saturation, after which m becomes negative. The inversion of m is extremely significant for its persistence and uniformity, showing a very interesting property; that is, the liquid dipoles change their orientation immediately after saturation, occupying an antimagnetization stand, that is, in opposition to the external magnetic field.

The other transport property χ is the magnetic susceptibility whose mode of change with respect to the applied field is significant, and, interest-

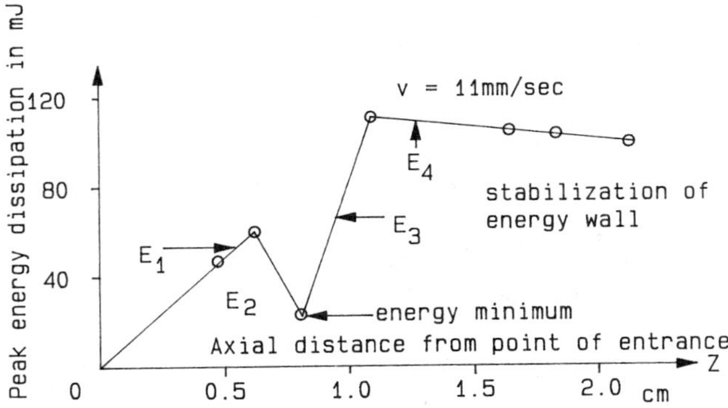

Figure 4.10 Energy representation of domain walls in ferric solution. ("Magnetic Transport Properties of Dipolar Conducting Liquids at Ambient Temperature by K. Denno for Journal of Electrostatics, 7, 1979. © Elsevier Science Publishers, Amsterdam, 1979.)

ingly, χ is a measure for the ratio of the magnetization m or magnetic induction B with respect to the applied magnetic field H_a. The magnetic susceptibility surges to an impulse of opposite values exactly at saturation. It is a solid proof that saturation is total and that both the magnetic dipole moment per unit volume and magnetic flux density B are at their peak levels.

To establish a correlation for the domain walls' energy distribution limit along the fluid continuum, a plot for the peak gyromagnetic energy with respect to the location of any point shown in Figure 4.10 is extracted directly from a similar set of measurements of the gyromagnetic power versus the fluid linear velocities.

4.11 PARAMETRIC MODEL OF THE REDOX CELL[23]

We may predict from the preceding experimental and theoretical analysis that the equivalent gyromagnetic resistance R_{gm} is a function of the magnetic field exciting current and the linear velocity as well as the concentration of the ferric solution in the catholyte of the redox flow cell. Also with reference to Lenz's law, there ought to exist an equivalent self-inductance for the time-changing magnetic field within the catholyte continuum. This time-changing field could be attributed to the reorientation of the ion magnetic dipoles or magnetites that results in a counterfield which is established by the induced current i_{gm} (the induced gyromagnetic current).

Therefore from the above reasoning, and taking into consideration the

[23] © 1976 IEEE. Reprinted, with permission, from Proceedings—IEEE Communication and Power Conference, Oct. 1976 pp. 403–406, Montreal, Canada. Paper entitled: "Modelling of Redox Flow Cell, Fuel Cell, Storage Battery and Harmonically Commutated Inverter" by K. Denno.

capacitive effect of the redox flow cell simulating its energy storing function, the parametric representation for the redox-flow-cell equivalent circuit is that shown in Figure 4.11, where as the resistive, inductive, as well as capacitive roles are represented.

Now, we are at a stage where it is logical and important to establish a dynamic model for the redox flow cell in the complex frequency domain which could be used as a tool for solving the time response of this energy-producing device as well as the steady-state performance.

Turning to Figure 4.11,

$$i = i_c = C\frac{dE_c}{dt} \tag{4.47}$$

$$i = i_L = \frac{1}{L_{gm}}\int E_L dt \tag{4.48}$$

$$i = i_{gm} = \frac{E_R}{R_{gm}} \tag{4.49}$$

Therefore,

$$E = \frac{1}{C_{gm}}\int idt + iR_{gm} + L_{gm}\frac{di}{dt} \tag{4.50}$$

Taking the Laplace transform of equation 4.50 and assuming zero initial conditions results in the following relationship:

$$\frac{E(S)}{S} = \frac{I(S)}{SC_{gm}} + \sum_{k=1}^{P}\frac{A_1(S_k)}{B_i(S_k)}\, r_{gm}(S - S_k) + SL_{gm}I(S) \tag{4.51}$$

where

$$k = \text{the pole order}$$

$$I(S) = \frac{A_1(S_k)}{B_1(S_k)} \tag{4.52}$$

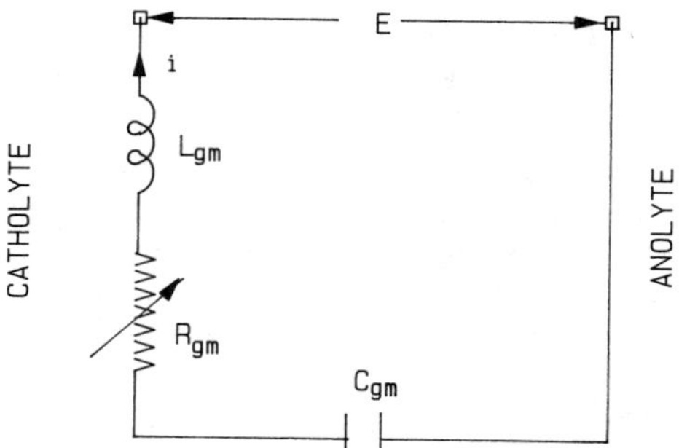

Figure 4.11 Parametric model for the redox flow cell.

$$r_{gm}(s) = \frac{A_2(S)}{B_2(S)} \tag{4.53}$$

$$L_{gm} = \text{independent of time}$$

$$E(t) = EU_{-1}(t), \text{ a step function}$$

From equations 4.51–4.53,

$$\frac{I(S)}{E(S)} = \frac{C_{gm}}{1 + CL_{gm}S^2} - \frac{C_{gm}}{E} \frac{S \sum_{k=1}^{P} [A_1(s)/B_i(s)] r_{gm}(S - S_k)}{1 + CL_{gm}S^2} \tag{4.54}$$

Let

$$G = C_{gm} \tag{4.55}$$

$$H = S^2 L_{gm} \tag{4.56}$$

Therefore,

$$\frac{I(S)}{E(S)} = \frac{G}{1 + GH} - \frac{C_{gm}}{E} \frac{\sum_{k=1}^{P} [A_1(s)/B_i(s)] R_{gm}(S - S_k)}{1 + C_{gm}L_{gm}S^2} = Y(s) \tag{4.57}$$

Then from equation 4.57, we can establish the dynamic model of the redox flow cell as shown in Figure 4.12, which is the admittance model in the complex frequency domain.

The admittance representation of the redox flow cell in the frequency domain points to the roles of R_{gm}, L_{gm}, and C_{gm} in controlling the cell performance in the transient as well as the steady durations.

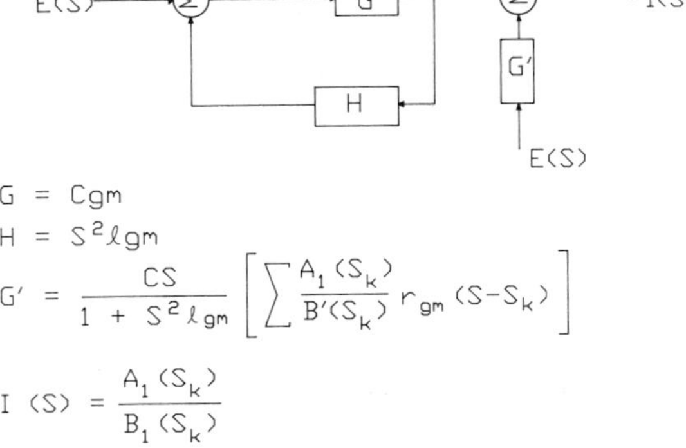

Figure 4.12 Dynamic model for the redox flow cell. ("Modelling of Redox Flow Cell, Fuel Cell, Storage Battery, and Harmonically Commutated Inverter" by K. Denno for IEEE Canadian Communications and Power Conference 1976 IEEE. Reprinted with permission.)

4.12 SUMMARY

In this chapter feasibility has been presented regarding design, cost analysis, as well as theoretical and experimental characterization of the redox-flow-cell system as an alternative energy source.

The following aspects of the redox-flow-cell power system are discussed:

First considered was the physical concept of the redox flow cell, that the device operates on the principle of oxidation-reduction whereby the ions of the redox couple remain soluble in their electrolytes in either oxidized or reduced states. System description of this composite as an alternative energy source is presented which centered on the redox couple $|TiCl_3\text{-}TiCl_4|$ $|FeCl_3\text{-}FeCl_2|$. The redox cell is rechargeable, and reversing the direction of flow of the electric current electrical energy is stored in chemical form. The two-tank redox flow cell was identified in Figure 4.1.

Then analytical treatment for the phenomena of magnetization and relaxation occurring inside the redox-flow-cell catholyte chamber between the dipolar ferric solution and the existing internal magnetic field was discussed, emphasizing the simplified concept of classical and quantum mechanics.

For the catholyte solution as a continuum of loosely floating magnetic dipole clusters, the magnetization is considered approximately equal to $N\mu'^2H/3kT$ and the density of energy transfer with respect to each dipole is approximately equal to $N\mu'^2H^2/3kT$ joules/m^3.

Next we saw that experimental simulation for the magnetofluid interaction in the catholyte chamber of the redox flow cell led to direct knowledge of correlation among the gyromagnetic energy, the time rate of magnetization, and the induced gyromagnetic current with respect to a DC exciting electric current in a coil surrounding the catholyte chamber, at various levels of catholyte solution concentrations.

Finally, theoretical analysis of transport properties of the catholyte fluid flow using the principles of paramagnetism and ferromagnetism (for loosely floating cluster dipoles) led to identification of the occurrence of magnetic resonance as well as energy distribution per domain walls throughout the catholyte continuum.

4.13 SOLVED EXAMPLES

1. If the current density in a 10-MW peak capacity redox flow cell ranges from 10.8 to 108 mA/cm^2, and the cell nominal voltage is 0.625 V, write an equation for power versus load current if the no-load power is $0.64A$, where A is the electrode area. Also calculate the number of cells connected in series to generate an output voltage of 25 kV and the electrode area of each cell.

Solution From data already presented,

$$\text{maximum current density} = 108 \times 10^{-3} \times 10^4$$
$$= 1080 \text{ amp/m}^2$$
$$\text{maximum power density} = 1080 \times 0.6$$
$$= 648 \text{ watts/m}^2$$
$$\text{minimum current density} = 10.8 \times 10^{-3} \times 10^4$$
$$= 1080 \text{ amp/m}^2$$
$$\text{minimum power density} = 108 \times 0.6$$
$$= 64.8 \text{ watts/m}^2$$

Then assume that the power density changes linearly with the load current i_L. If the minimum and maximum power densities are 64.8 and 648 W/m^2 at no load and full load, respectively, establish the straight-line relationship for power density $P_D(i_L)$, if the no-load value is 0.64 W/m^2.

$$P_D = ai_L + b \tag{1.a}$$

at $i_L = 0$, $P_D = 64.8$ W/m^2
and for

$$i_L = 1080 \text{ A/m}^2, P_D = 648 \text{ W/m}^2 \tag{1.b}$$

Therefore,

$$64.8 = 1080a + b$$
$$648 = 1080a + b$$
$$972a = 583.2$$
$$a = \frac{583.2}{972} = 0.6$$
$$b = 0$$

Therefore,

$$P_D = 0.625i_L + 0.64 \tag{1.c}$$

$$\text{number of cells connected in series} = \frac{25{,}000}{0.625}$$
$$= 40{,}000 \text{ cells}$$
$$\text{peak power/cell} = \frac{10^7}{40{,}000} = 250 \text{ watts}$$
$$\text{electrode area/cell} = \frac{250}{0.625 \times 648} = 0.617 \text{ m}^2$$

2. From equation 4.7, the threshold limit for the redox-flow-cell potential is expressed by $V_o = \Delta G/nF$. Discuss the status of electron-mole value if the membrane is changed from Cl^- to H^+.

Solution For Cl^- membrane according to the reduction equation 4.2, $n = 1$. Now referring to Figure 4.2, using H^+ membrane, the discharge reaction equation is

$$Ti^{+3} \rightarrow Ti^{+4} + e^-$$

n is also equal to 1.

Hence $\Delta G = 57.9 \times 10^6$ joule/kg-mole, which is the same as in the case of H^+ membrane, since hydrogen and chloride each have the valency of unity.

3. In the catholyte of the redox flow cell, if the externally applied magnetic field through a solenoid produces magnetic flux density of 2 tesla, find the effective magnetic moment required for a total energy absorption of 1 joule/m^3 at the ambient temperature of 25°C. Assume ion concentration $N = 60.2 \times 10^{26}/\text{m}^3$.

Solution From equation 4.15

$$E = \frac{N\mu_{\text{eff}}^2}{3kT} H^2$$

$$H = \frac{B}{\mu_o} \tag{3.a}$$

$$\mu_{\text{eff}} = \frac{3kTE}{NH^2}$$

Therefore

$$\mu_{\text{eff}} = \frac{3kTE\mu_o^2}{NB^2} \tag{3.b}$$

where

$\mu_o = 4\pi \times 10^{-7}$ henry/m, permeability of nonmagnetic continuum

Therefore

$$\mu_{\text{eff}} = \frac{3 \times 1.38 \times 10^{-23} \times 298 \times 1(4\pi \times 10^{-7})^2}{6.02 \times 10^{26} \times 4}$$

$$= 8.12 \times 10^{-60} \text{ joule/ampere-turns}$$

4. Refer to equation 4.21. If $H(t)$ is an impulse and $T(t)$ is a step function in the time domain, representing sudden surge in temperature, solve for the state of magnetization $M(t)$ with zero initial conditions.

Solution From equations 4.14 and 4.16,

$$\frac{dM}{dt} + \frac{2K}{\nu} T(t) = RH(t) \tag{4.a}$$

$$H(t) = H_o U_o(t)$$

$$\frac{dM}{dt} + \frac{2kT}{\nu} U_{-1}(t) = RH_oU_o(t) \tag{4.b}$$

$$\frac{dM}{dt} = RH_oU_o(t) - \frac{2kTU_{-1}(t)}{\nu} \tag{4.c}$$

Now, integrate both sides of 4.c to obtain the following:

$$M(t) = Rh_o - \frac{2kT}{\nu} t + \text{zero}$$

where R is a function of position and independent of time.

5. Refer to Figures 4.7.a and 4.7.b to identify the transformation from a state of magnetization to relaxation and vice versa with respect to the concentration of moving ferric solution. Discuss data.

Solution From Figure 4.7.a,

velocity = 1 cm/sec		velocity = 2 cm/sec	
R_{gm} (inversion)	Concentration	R_{gm} (inversion)	Concentration
600	1%	450	2%
500	2%		
200	6%		

We can see that at a slower motion (1 cm/sec) of ferric solution the frequency of inversion is more than at 2 cm/sec, which is expected, since magnetites at 1 cm/sec have more time of exposure to magnetic field effect. Also we notice that at 1 cm/sec, ohmic resistance inversion will diminish at a higher concentration, where at 6 percent, it is 200Ω compared to 600Ω and 500Ω at 1 percent and 2 percent, respectively, which could be attributed to the viscosity effect and also with reduced influence from magnetic field effect.

6. Express the gyromagnetic energy W_{gm} expended during magnetic alignment of magnetites in terms of local particle effect.

Solution Treating the ferric solution as a ferromagnetic fluid with loosely floating magnetites, W_{gm} expressed in equation 4.43 states that

$$W_{gm} = N\frac{\mu^2H^2}{3kT} \tag{6.a}$$

H is the total magnetic field, where

$$\overline{H} = \overline{H}_a + \gamma\overline{M} \tag{6.b}$$

H_a = the applied magnetic field intensity in At/m
γ = Weiss constant
M = the magnetization or dipole moments/meter^3

Hence,

$$W_{gm} = \frac{N\mu'^2}{3kT}(H_a + \gamma M)^2 \tag{6.c}$$

$$= \frac{N\mu'^2}{3kT}(H_a^2 + 2\gamma M H_a + \gamma^2 M^2) \tag{6.d}$$

Therefore,

$$W_{gm} = \frac{N\mu'^2 H_a^2}{3kT} + \frac{2\gamma M N \mu^2 H_a^2}{3kT} + \frac{N\mu^2\gamma^2 M^2}{3kT} \tag{6.e}$$

Equation 6.3 identified three phenomenological aspects to the energy of magnetization, namely, on the right-hand side. The first term is due to the applied field H_a. The second term is due to the interaction between the local internal field and the external field. The third term is due to internal or molecular existing field.

7. Referring to Figure 4.9.a, express the ferric solution permeability at the field boundary of magnetic inversion.

Solution From equation 4.31

$$\mu = 1 + 4\pi\chi \tag{7.a}$$

The boundary of magnetic inversion, which is from Figure 4.9.a, occurs when $H = 0.9$ At/m

$$\chi^+ = 64 = 10^{-6}\, U_o(H - 0.9) \tag{7.b}$$

and

$$\chi^- = -70 \times 10^{-6}\, U_o(H - 0.9) \tag{7.c}$$

where $U_o(H - 0.9)$ is a delayed impulse
Therefore, before inversion

$$\mu^+ = 1 + 4\pi \times 64 \times 10^{-6}\, U_o(H - 0.9) \text{ henry/m} \tag{7.d}$$

and just after inversion

$$\mu^- = 1 - 4 \times 70 \times 10^{-6}\, U_o(H - 0.9) \text{ henry/m} \tag{7.e}$$

8. Refer to equation 4.57 describing the dynamic model of the redox flow cell in the frequency domain. Find the admittance model in the steady state. The redox cell model in the steady state implies that $s \to 0$, where according to the final value theorem,

$$\lim_{s\to 0} SF(s) = \lim_{t\to\infty} f(t) \tag{8.a}$$

Therefore

$$\lim_{s\to 0} S\,\frac{I(S)}{E(S)} = \lim_{t\to\infty} Y(t) \tag{8.b}$$

$$= \lim_{s \to 0} \left[\frac{SC_{gm}}{(1 + S^2 L_{gm} C_{gm})} - \frac{C_{gm} S^2}{E} \frac{\sum_{k=1}^{P} [A_1(s)/B_i(s)]/R_{gm}(S - S_k)}{(1 + C_{gm} L_{gm} C^2)} \right] \quad (8.c)$$

$$= \lim_{s \to 0} \left[\frac{SC_{gm}}{(1 + S^2 L_{gm} C_{gm})} - \frac{C_{gm}}{E} \frac{\sum_{k=1}^{P} [A_1(s)/B_i(s)]/R_{gm}(S - S_k)}{(1/S^2 + C_{gm} L_{gm})} \right] \quad (8.d)$$

$$= 0$$

on the assumption that $[A_1(s)/B_i'(s)]R_{gm}(S - S_k)$ is independent of (s) (a special case).

4.14 REVIEW QUESTIONS

1. List major problems that may adversely affect the efficiency of the redox flow cell.
2. Identify the basic electrochemical problem that negatively affects sustained operation of the redox flow cell as a reliable power plant.
3. Identify the basic electromagnetic problem that must be resolved to improve drastically the redox-flow-cell efficiency.
4. Discuss the role of the catholyte pumping fluid velocity and concentration on the power dissipation phenomenon in that chamber.
5. Identify the physical implication on the gyromagnetic resistance R_{gm} as the current source of the applied magnetic field increases.
6. Refer to Figure 4.5a–c. Explain the asymptotic pattern of W_{gm} with respect to i_L.
7. Correlate the occurrence of maximum P_{gm} in Figure 4.6.a–e and that of magnetic inversion for m in Figure 4.9.a–e.
8. Compare what you see in Figure 4.10 regarding the patterned behavior of magnetization energy distribution in ferric fluids with that of the Bloch walls pattern in ferromagnetic solids.
9. Referring to Figure 4.9.a and b, comment on the occurrence of magnetic inversion at a specific external magnetic field at the two different fluid motions.
10. Convert the voltage output model of the redox flow cell (shown in Figure 4.11) to a current output model.

4.15 PROBLEMS

1. Referring to Figure 4.4 introduce the overall chemical reaction equation describing the process of recharging of the redox flow cell.
2. Given a 10-MW lithium-iron redox-flow-cell system, with 65 percent efficiency, electrode current density of 50 ma/cm^2, and no-load terminal voltage per cell of

0.8 volt, calculate the number of redox cells to be connected in series-parallel combination.

3. A titanium-iron redox flow cell has its terminal output voltage approximately linear and expressed by $U_t = A - iB$, where A and B are constants. If the electrode area is Am^2, find the current density for maximum power output.

4. If the output no-load terminal voltage for a redox-flow-cell interconnected system is 1 kV and total change in Gibbs free energy per cell is 58×10^6 joule/kg-mole, find the total number of electron moles associated with the total reactions.

5. In the catholyte of a redox flow cell, the pumping velocity of the ferric solution is 2 cm/sec and the behavior of m as well as the magnetization energy are as shown in Figure 4.9.a. Obtain the

a. Solution for μ, the magnetic permeability

b. Functional form for χ

c. Functional form for the magnetization energy W_{gm}

6. In the catholyte of a redox flow cell, establish a mathematical expression with a plot for the magnetic susceptibility U_s pumping velocity of the ferric solution. Refer to Figure 4.9.a through e at the threshold of inversion.

7. Repeat Problem 6, but for the magnetic dipole moment/unit volume and the magnetic (m) susceptibility U_s pumping velocity of the ferric solution.

8. Establish a mathematical correlation in the form of a plot for the gyromagnetic power in the catholyte of the redox flow cell versus ferric solution percentage concentration at the threshold of maximum W_{gm}. Refer to Figure 4.6.a through e for ferric solution velocity of 1 cm/sec.

9. Repeat Problem 8, but for ferric solution pumping velocity of 2 cm/sec. Compare this pattern with that of Problem 8.

10. Predict adverse implications of the expended gyromagnetic power in the catholyte on the performance efficiency of the redox flow cell.

11. Refer to Figure 4.7.a and b, and establish the appropriate mathematical singularity function describing the state of inversion in $R_{gm}U_s$ at a concentration for a ferric solution velocity of 1 cm/sec and then 2 cm/sec.

12. Refer to Figure 4.8.a and b, and establish the appropriate mathematical singularity function of R_{gm} for peak $P_{gm}U_s$ concentration of ferric solution pumping velocity of 1 cm/sec and then 2 cm/sec.

13. Refer to equation 4.22 describing space and time differentials for the state of magnetization and relaxation. Given that m is a step function and H is a single square pulse in the time domain, solve for the magnetite space distribution function $F(x)$.

14. Repeat Problem 13 if $m(t) = Ae^{-t/\tau}$, where τ is the time constant for magnetization and $H(t)$ is an applied sustained DC field, that is, a step function.

15. Repeat Problem 13, if $m(t)$ and $H(t) = H_o(\sin \omega t + \sin 3\omega t)$ with zero spatial or time conditions. $U_{-1}(t)$ is a step function.

16. Establish a new mathematical model in the (S) domain for the redox flow cell at the state of resonance where $SL_{gm} = 1/SC_{gm}$ and

$$I(S) = \frac{W}{S^2 + W^2}, \qquad R_g(S) = \frac{R}{S}$$

17. With the occurrence of a sudden short circuit at the terminals of the redox flow cell, establish its mathematical model in the (S) domain and general solution of the short-circuiting current in the time domain. You can assume arbitrary independent function at integration.
18. Convert the redox-flow-cell parametric circuit model shown in Figure 4.11 into a parallel of equivalents of R, L, and C. In the complex frequency (S)-domain, identify the bandwidth as well as the upper and lower half-power frequencies.
19. The Langevin function shown in Fig. P.1 has an initial slope of 1/3, followed by a curved path similar to the normal magnetization curve (B-H) of ferromagnetic reactors. Establish at least two new functional forms for the magnetization m of equation 4.8 based on an empirical form of $L(x)$. Compare the new two functional forms with that of equation 4.9.

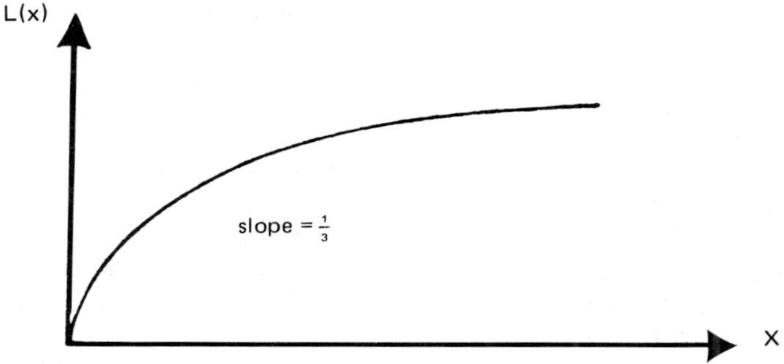

Figure P.1. Langevin function L, (x).

20. The magnetization-relaxation equation for a water-based magnetite solution is expressed in equation 4.14. If $F(x)$ is an impulse function in space and $H(t)$ is a DC field, solve for $m(t)$ in terms of all parameters in this equation.
21. Refer to the dynamic model for the redox flow cell shown in Figure 4.12. Given $I(S) = I_o/S^2$ and $R_{gm}(S) = R$, and using the principles of network synthesis, establish an overall equivalent network for the redox flow cell in terms of physically realizable elements.

4.16 REFERENCES

1. Ateya, B. G., and L. G. Austin. "The Kinetics of $Fe^{2+}/FeCl_2/HCl_2$ (AG) on Pyrolytic Graphite Electrodes." *Electrochemical Science and Technology,* September 1973.
2. Bogardus, H., D. A. Krueger, and D. Thompson. "Dynamic Magnetization in Ferrofluids." *Journal of Applied Physics,* June 1978, pp. 3422–3429.
3. Cobine, J. D. *Gaseous Conductors.* New York: Dover Publications, 1958.

4. Dekker, A. J. *Solid State Physics.* Englewood Cliffs, N.J.: Prentice Hall, 1963.
5. Denno, K. "Magnetic Transport Properties of Dipolar Conducting Liquids at Ambient Temperature." *Journal of Electrostatics,* Vol. 7 (1979), pp. 345–360.
6. Denno, K. "Modelling of Redox Flow Cell, Fuel Cell, Storage Battery and Harmonically Commutated Inverter." *Proceedings of the 1976 Canadian Communications Power Conference,* Montreal, Canada, pp. 403–406.
7. MIT Staff. *Magnetic Circuits and Transformers.* New York: John Wiley & Sons, 1947.
8. Peterson, E. A. "Reversible, Field Induced Agglomeration in Magnetic Colloids." *Journal of Colloid and Interface Science,* Vol. 62, no. 1 (October 1977), pp. 24–34.
9. Rabinowitch, E., and W. H. Stockmeyer. "Association of Ferric Ions with Chloride Bromide and Hydroxyl Ions." Contribution No. 482 from the Research Laboratory of Physical Chemistry and Publication No. 9, Solar Energy Conversion Research Project, MIT, Cambridge, Mass., 1942.
10. Smity, A. C., J. F. Janak, and R. B. Adler. *Electronic Conduction in Solids.* New York: McGraw-Hill Book Co., 1967.
11. Speck, C. E., J. J. Lee, and O. K. Mawardi. "Interaction of a Plasma Beam with a Transverse Magnetic Barrier." *IEEE Transactions on Plasma Science,* September 1977.
12. Solverman, J., and R. W. Dodson. "The Exchange Reaction Between the Two Oxidation States of Iron in Acid Solution." Research Report of the Chemistry Department, Columbia University and Brookhaven National Laboratory, Upton, New York, 1952.
13. Thaller, Lawrence. "Electrically Rechargeable Redox Flow Cells." NASA TM-X-71540, Lewis Research Center, Cleveland, Ohio, August 1974.
14. Walsh, E. M. *Energy Conversion.* New York: The Ronald Press, 1967.
15. Warshay, Marvin, and Lyle O. Wright. "Cost and Size Estimates for an Electrochemical Bulk Energy Storage Concept." NASA TM-X-3192, Lewis Research Center, Cleveland, Ohio, 1975.
16. Woodson, H. H., and J. R. Melcher. *Electromechanical Dynamics,* Vols. I, II and III. New York: John Wiley & Sons, 1968.

5

Dynamic Modeling of Basic Types of Solid State Power Inverters

5.1 INTRODUCTION

In anticipation of the development of a power system containing a combination of conventional electromechanical buses and new static electrochemical, solar power, redox flow power, biomass power, and any other mode of alternative energy busbars, a systematic approach is being proposed for modeling the various components of the static system generating buses.

The electrochemical bus, for example, could be the combination of either a fuel-cell–inverter–filter–transformer system or a storage battery–inverter–filter–transformer system.

This chapter concerns the mathematical modeling of several types of solid-state inverters with respect to their mode of commutation. The function of the inverter is to transfer a direct current source to an acceptable AC output, where a wave shape can be established by a set of selenium-cadmium rectifiers (SCRs) to form rectangular pulses of varying widths and amplitudes and regularly reversing polarity. These pulses can be combined to form stepped waves resembling sine waves, and then by proper filtering, a pure fundamental sine wave can be obtained.

The SCR in the inverter blocks current flow in the forward direction until a trigger signal is applied to its gate where timing will commence at the beginning of the pulse. After turning the SCR on, however, it is somewhat difficult to turn it off properly at the required instant.

The actual transfer of current from one SCR to another in the process of inversion is called commutation, which involves the following steps:

1. The complete decay of a current to zero in one SCR.
2. The reapplication of a forward voltage to this SCR after regaining its forward blocking capability.
3. Transfer of current to the next SCR in sequence.

A three-phase inverter can be structured by combining three basic single-phase inverters, whose output transformer secondaries may be connected in either a Y or a Δ system to supply a three-phase load.

Previous work in connection with basic inverters indicates that its physical size, rating, and losses depend to a large extent on the pulsating nature of DC current drawn from the source and the degree of its voltage fluctuation. Switching operation of the inverter can cause harmful transient overvoltage that may be amplified more seriously by the internal inductance of the DC source and even that of the cable connecting the DC source to the inverter. It is also established that multiphase operation of inverters has the effect of minimizing the adverse action of ripples in the DC source pulsating current.

In moving toward complete and useful utilization of static generating buses in power systems, steady-state and dynamic studies of the complete integrated system are needed, since power systems involve a combination of rotary or electromechanical and static power generating busbars.

As a first step toward dynamic solutions of the above-mentioned integrated power systems, dynamic modeling of various basic types of solid-state power inverters is presented here. This will be followed by modified types of inverters with respect to their improved method of commutation. After establishing the dynamic modeling of the DC sources, which could be either storage battery or fuel cell, a complete model of the whole static generating bus could be structured.

Previous research in the field of inverters confined itself to its application for supplying variable frequency power to variable speed drives, including the performance of various methods of commutation, component values, and significance of losses. Performance calculation of parallel-capacitor–type commutation inverters at various frequency ranges were investigated at length, indicating that inverter output at power frequency range is feasible with moderate value of the commutating capacitor. Also, it had been established that the most efficient mode of operation of this parallel-capacitor–type inverter gives rectangular wave output similar to that of inverters feeding infinite busbars.

Other work includes the effective means of achieving output voltage regulation within the inverter through a scheme of harmonics neutralization, leading to improvement in efficiency and reduction in losses.

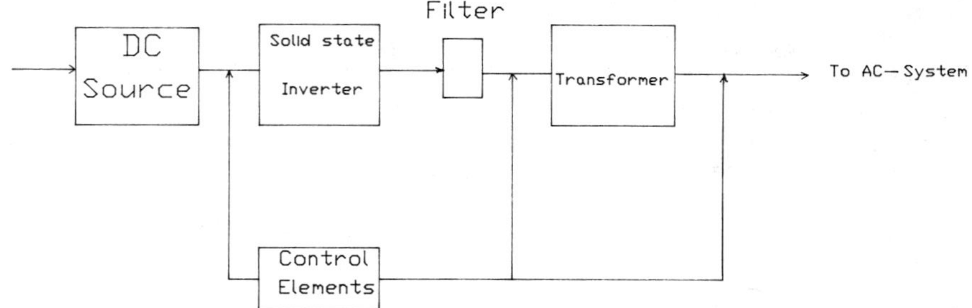

Figure 5.1 Location of the inverter filter within static electrochemical power busbar.

Analysis of various modes of inverter commutation has been presented emphasizing the significance of operation of each mode toward perfecting the act of diverting current flow of a conducting path from one SCR to another.

5.2 OBJECTIVES

The main objective aims in this chapter are the development of dynamic modeling of a polyphase solid-state inverter used for bulk power generation. The modeling will be simulated in terms of mathematical representation in the complex frequency domain that will enable us to secure the steady-state and transient performance of any mode of inverter.

Location of the inverter at the static electrochemical generating bus, for example, is shown in Figure 5.1. Models of several basic types of inverters will be presented throughout this chapter text and in solved examples, each representing basic mode of commutation:

Direct-line commutated inverter
Parallel-capacitor forced commutated inverter
Series-capacitor forced commutated inverter (type 1)
Series-capacitor forced commutated inverter (type 2)
Transformer forced commutated inverter
Auxiliary-impulse forced commutated inverter
n-Phase externally commutated inverter
n-Phase internally commutated inverter

5.3 MODELING OF INVERTERS

Modeling of solid-state inverters in the complex frequency domain is essential for the ultimate goal of securing their steady-state and transient response subject to a variety of forcing actions. The importance of modeling of inverters is reflected by the following striking conditions:[1]

1. Complexity of the inverter solid-state circuit as well as the interconnected filter system
2. The imposed necessity of the control feedback system required to regulate the inverter-filter sinusoidal output
3. The system arrangement of the loading system, whether multiload, nonlinearity of the load, pulsing of load, and switching sequence of loads
4. Effect of source DC voltage, impedance variations of the inverter-filter system, as well as stability of the solid-state SCRs

In this chapter the process of circuit and dynamic modeling will be presented for the direct-line commutated inverter, the parallel-capacitor inverter, the transformer commutated inverter, the *n*-phase externally commutated inverter, and the *n*-phase internally commutated inverter.

5.3.1 Modeling of Direct-Line Commutated Inverter[2]

Circuit Diagram and Operation

The actual rectifier circuit (see Fig. 5.2) can be converted to function as inverter by making two basic design modifications:

1. The two terminal rectifying elements must be replaced with controlled rectifier devices.
2. Adequate means of commutation must be provided.

With gradual variation in the firing angle of the phase-controlled rectifer, the circuit can function in the inverse operation with adequate reversal of power.

Commutation is achieved automatically by the time relationship existing among the AC line voltages. Power will flow from the DC source to the

[1] S. R. Bowens, J. Clare and R. R. Clements. "Transient Performance of Inverter Systems." © 1982 IEE. Reprinted, with permission, from IEE Proceedings, Vol. 289, Pt-B, pp. 301–303, November 1982.

[2] Ibid. and K. Denno "Dynamic Modelling of Basic Types of Solid State Power Inverters" in paper presented at the 1976 IEEE Midwest Power Symposium, October 1976. Also from B. D. Bedford and R. C. Holt, *Principles of Inverter Circuits*. (New York: John Wiley & Sons, Inc. 1964).

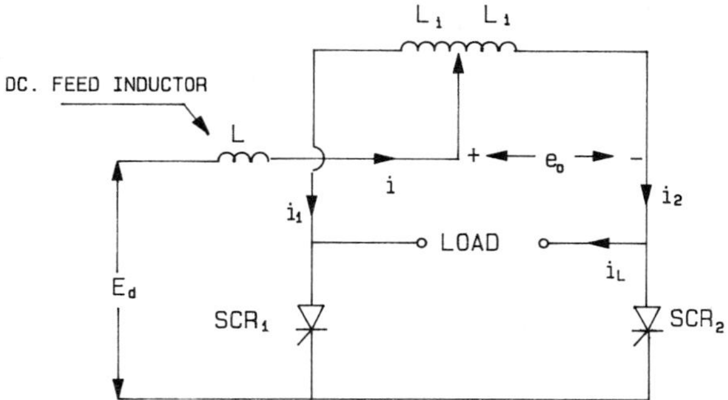

Figure 5.2 Circuit diagram of the direct-line commutated inverter. (from B. D. Bedford and R. C. Holt, *Principles of Inverter Circuits*. New York: John Wiley & Sons, Inc. 1964.)

AC line only if the SCRs are fired at a certain time relationship between amplitude and phase.

SCR_2 is fired when the AC line generates the polarity indicated and $E_d > e_o$. Limitation of this type of inverter is attributed to the case where the kilovolt-amp load (KVA) is greater than the inverter rating.

Dynamic Model

Based on the principled mode of the commutation process, and the fact that current in each SCR must be zero prior to the end of each conducting period, for this process we can proceed to write a set of circuit equations in the time domain as indicated.

Referring again to Fig. 5.2 we can write the following basic relationships:

$$i_2 = i_L \text{ SCR}_1 \text{ is conducting; SCR}_2 \text{ is open)}$$

$$E_d = \frac{V_L}{2} + L\,\frac{di}{dt}$$

$$L_1\,\frac{di_1}{dt} = V_L + L_1\,\frac{di_2}{dt} \tag{5.1}$$

$$i = i_1 + i_2$$

$$i = 0 \text{ prior to end of SCR conducting period}$$

$$i_1(t_o),\ i_2(t_o) = \text{initial values of } i_1 \text{ and } i_2 \text{ for the new conducting period}$$

$$V_L(S) = \text{transform of load voltage}$$

$$I_L(S) = \text{transform of load current}$$

Based on the components of equation 5.2, a dynamic model in the form of block diagram in the complex frequency domain is shown in Fig. 5.3.

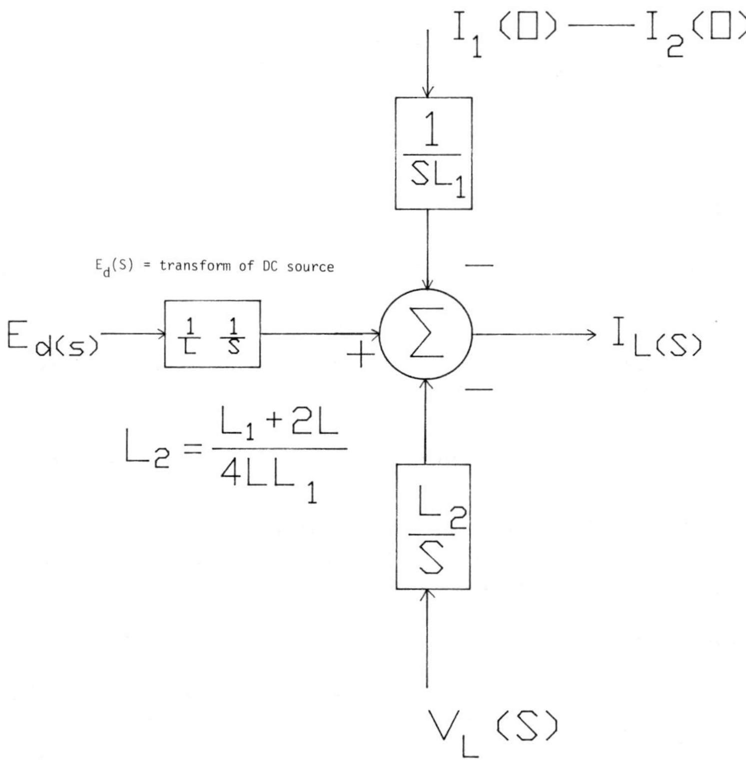

Figure 5.3 Dynamic model for the direct-line commutated inverter. (© 1982 IEE, Bowens, Clare and Clements paper; also presented at the 1976 IEEE Midwest Power Symposium, October 1976, Kansas State University by K. Denno.)

Symmetry is assumed in subsequent half-cycle operation and commutation. Taking the Laplace transformation of equation 5.1 and recognizing the fact that the current in any SCR must vanish before the end of any conducting period, the following transform model equation is obtained:

$$\frac{I_L(S)}{E_d(S)} = \frac{1}{SL} - \frac{V_L(S)}{E_d(S)}\frac{L_2}{S} - \frac{[i_1(t_o) - i_2(t_o)]}{E_d(S)}\frac{1}{SL_1} \tag{5.2}$$

The three-phase design of the direct-line commutated inverter is the delta inverter, introduced by P. D. Evans, R. C. Dodson, and J. F. Eastham. This mode of inverter is a new developing device to produce three-phase AC output through operational performance of three power transistors as shown in Figure 5.4.

Comparing its structural components with the bridge inverter, the delta inverter normally required half of what is needed in the bridge inverter.

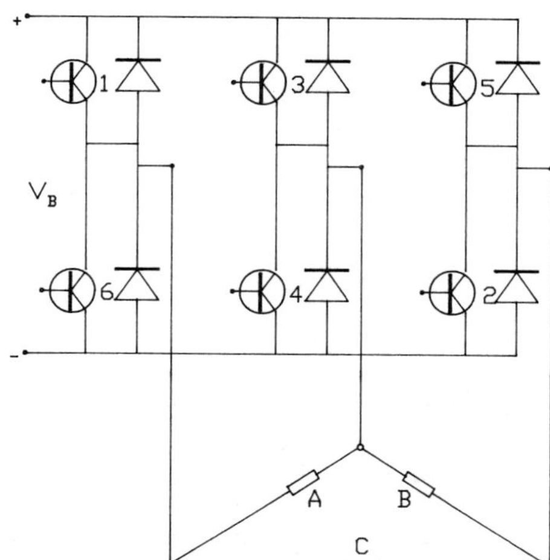

Figure 5.4 Bridge inverter and load. (© 1980 IEE. Reprinted with permission, from P. D. Evans, R. C. Dodson, & J. F. Eastham, "Delta Inverter.")

Therefore, reliability and cost-effectiveness of the three-phase delta inverter are superior to that of the bridge inverter as shown in Figure 5.5.

It is significant to present an equation for the peak voltage magnitude of the nth harmonic component, where

$$V_n = \frac{6E_d}{n} \sin \frac{2n}{3} \cos n \left(\frac{2\pi}{3} - \frac{\theta}{2}\right) \tag{5.3}$$

is the conduction angle.

5.3.2 Modeling of the Parallel-Capacitor Commutated Inverter

Circuit Diagram and Operation

The DC feed inductance forces the equivalent current waveform into the AC portion of the circuit and prevents excess current from passing into

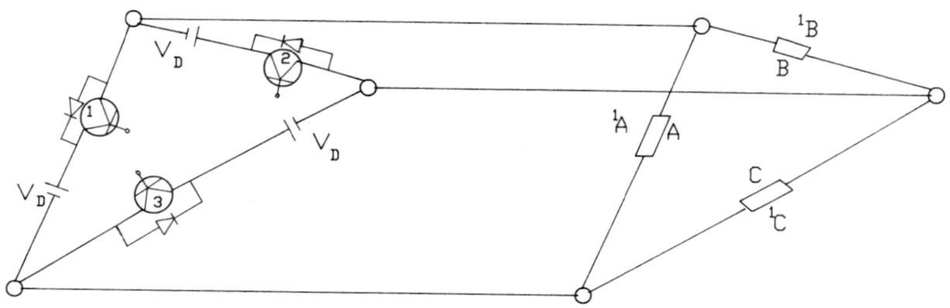

Figure 5.5 Delta inverter and load. (© 1980 IEE. Reprinted with permission, from P. D. Evans, R. C. Dodson, & J. F. Eastham, "Delta Inverter.")

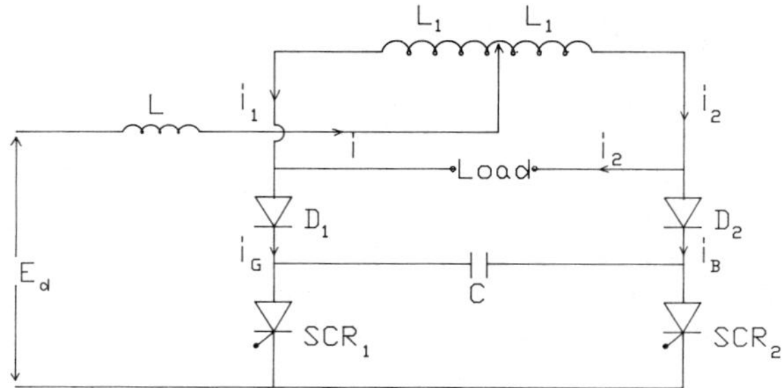

Figure 5.6 Circuit diagram for the parallel-capacitor commutated inverter. (© 1968 IEEE. Reprinted, with permission, from IEEE Transactions on Industry and General Applications, Vol. IGA-4, No. 1, Jan./Feb. 1968, pp. 104–110. Paper entitled: "Inverter Commutation Circuits" by Andrew J. Humphrey.)

the commutating capacitance during switching intervals (see Figure 5.6). A brief description of the commutation process follows.

The DC source sends current to the two halves of the stepping transformer, generating opposing magnetomotive forces (mmfs) and, with the DC feed inductance, prevents any cyclic variations in the DC current input.

With SCR_1 conducting, the capacitor will be charged to a certain voltage, and when SCR_2 is turned on, the capacitor voltage provides a reverse action, turning off SCR_1. A disadvantage of this type of inverter is that a large capacitance is needed to supply a large reactive load.

The general commutation performance could be described for each half-period by four distinct time intervals:

Interval 1. Begins when SCR_1 is triggered. The capacitor being previously charged extinguishes SCR_2 and then continues to decay.

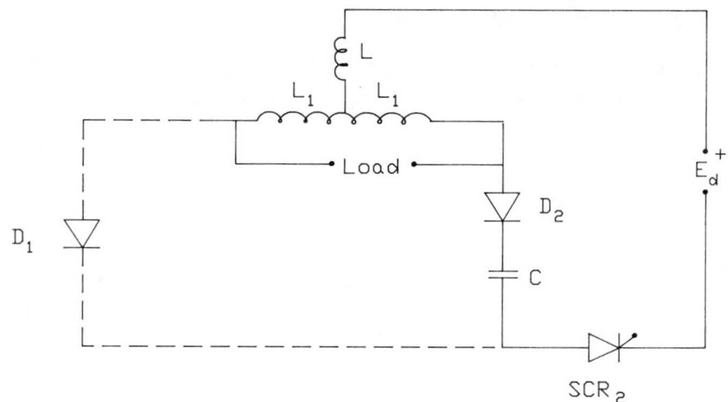

Figure 5.7.a The first commutation interval of the parallel-capacitor inverter. (All figure parts: © 1968 IEEE. Reprinted with permission from A. J. Humphrey, "Inverter Commutation Circuits." Also, © IEE, E. E. Ward, "Inverter Suitable for Operation Over a Range of Frequency.")

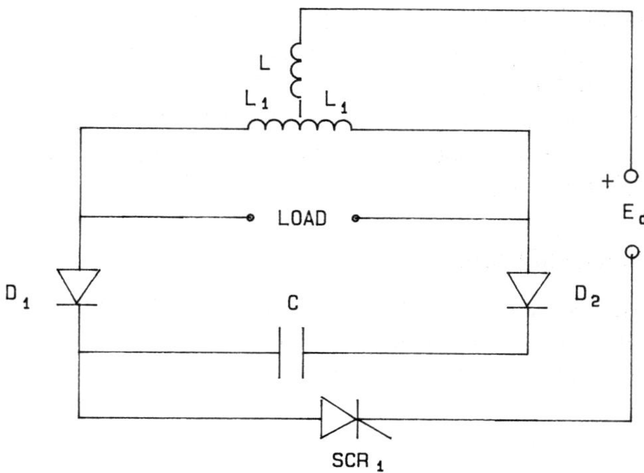

Figure 5.7.b The second commutating interval of the parallel-capacitor inverter.

This interval ends when the clamping diode D_2 starts to conduct (see Fig. 5.7.a).

Interval 2. As D_2 starts to conduct, the capacitor has decayed to zero. Inductors L and L_1 form with the load and the commutating capacitor C, a two-energy storage circuit during which the supply current generates a peak voltage, reversing the capacitor potential, where the peak capacitor current vanishes and consequently D_1 ceases to conduct (see Fig. 5.7.b).

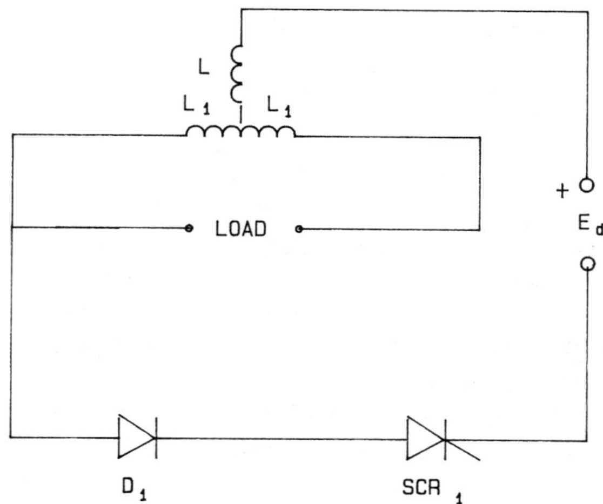

Figure 5.7.c The third commutating interval of the parallel-capacitor inverter.

Interval 3. Starts at peak voltage when D_1 ceases conducting thereby isolating the capacitor. This interval ends when load current ceases (see Fig. 5.7.c).

Interval 4. Load current and voltage are very small, DC power also slowly rises if DC feed inductance is large, and this period ends when SCR_2 is fired.

The Dynamic Model

Referring in Figs. 5.6, 5.7.a, 5.7.b, and 5.7.c, where the second interval of commutation is the major contributor for load voltage buildup, we can write the following circuit equations in the time domain:

$$
\begin{aligned}
E_d &= L\frac{di}{dt} + \frac{V_L}{2} \\
V_L &= \frac{1}{C}\int i_B dt = V_C \\
L_1\frac{di_1}{dt} + V_L &= L_1\frac{di_2}{dt} \qquad (5.4) \\
i &= i_1 + i_2 \\
i_B &= C\frac{dv_L}{dt}
\end{aligned}
$$

And, then, proceeding to take the Laplace transform of equation 5.4 and recognizing that electric current into any SCR must vanish prior to the end of the conducting period, the following mathematical representation in the complex frequency domain is secured:

$$
\frac{I_L(S)}{E_d(S)} = \frac{1}{2LS} - \frac{V_L(S)}{E_d(S)}\left(\frac{L_1 + 2L}{2LL_1}\right)\left(\frac{1}{S}\right)\left(1 + \frac{2LL_1C}{L_1 + 2L}S^2\right) + \frac{I_1(O)}{E_d(S)}\frac{1}{S} \qquad (5.5)
$$

As a follow-up for the parallel-capacitor commutated inverter introduced by Ward and Subrahmanyam, see the design of the current source inverter as shown in Fig. 5.9. Current source inverters are becoming increasingly preferred over voltage inverters where they are inherent in dealing with high inverse voltages.

Consequently, we can now proceed to construct the dynamic model of the parallel-capacitor commutated inverter in the complex frequency domain based on the components of equation 5.4. This model is shown in Fig. 5.8.

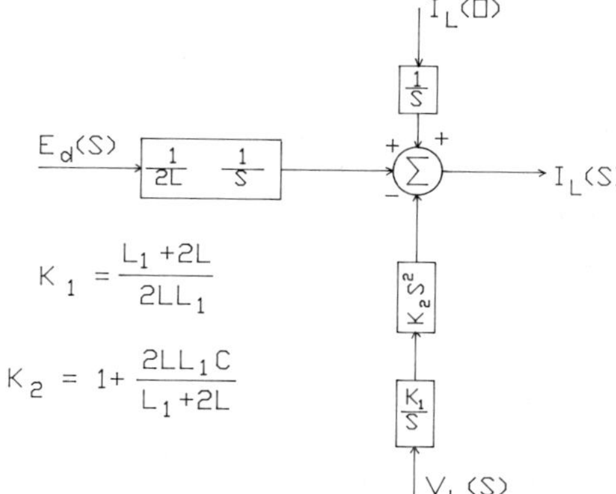

Figure 5.8 Dynamic model of the parallel-capacitor inverter. (Reprinted with permission of Energy Utilization of P. S. E. & G. of NJ; 1976 IEEE Midwest Power Symposium.)

5.3.3 Modeling of the Transformer Commutated Inverter

Circuit Diagram and Operation

We can present a brief description of the transformer commutated inverter which is voltage commutation with direct transfer of load current as shown in Figure 5.10. It has the advantage that power loss in the commutation process is being taken care of by the auxiliary circuit rather than by the direct inverter circuit where the trapped energy is mixed with the load current.

Load current prior to commutation process is passing through SCR_1, and when SCR is fired, the current (i) in the transformer-capacitance loop

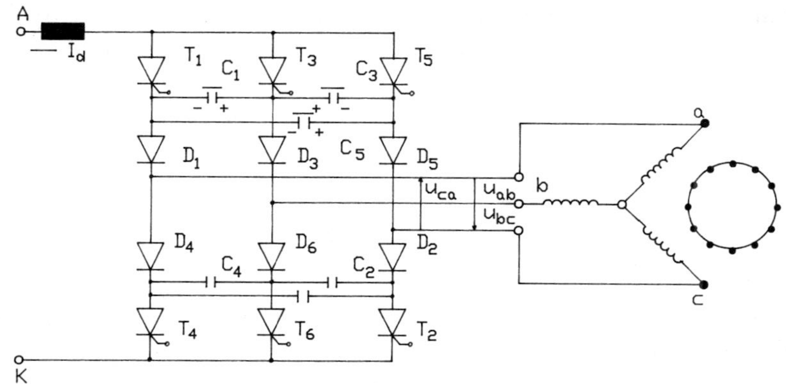

Figure 5.9 Current-source inverter feeding a three-phase induction motor. (© 1983 IEE. Reprinted, with permission, from the IEE Proceedings Vol. 130, Pt-B, No. 5, pp. 355–359, Sept. 1983. Paper entitled: "Analysis of Commutation of a Current Source Inverter Feeding an Induction Motor, Using 2-Axis Variables" by Vedam Subrahmanyam.)

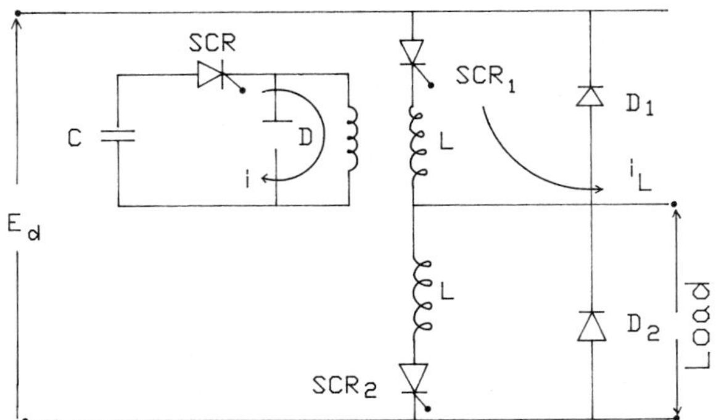

Figure 5.10 Circuit diagram of transformer commutated inverter. (© 1968 IEEE. Humphrey, "Inverter Commutation Circuits.)

will increase toward the limit of load current to match the principle of the constant linkage theorem.

The voltage across the capacitor will increase, and as soon as it becomes equal to or exceeds the supply voltage, the load current will shift to the path through D_2. The capacitor will discharge, storing energy into the transformer magnetizing reactance, and the SCR_1 will regain control as the reduced capacitor voltage approaches the supply voltage.

The stored energy into the transformer reactance will establish a reversed voltage, and hence the stored energy will be discharged through diode D.

Dynamic Model

Referring to Fig. 5.10 and considering the function of the process of commutation regarding the firing of the SCRs and storing capability of the transformer reactance, we can write the following circuit equations:

$$\frac{1}{C}\int i\,dt + L\frac{di}{dt} + M\frac{di_L}{dt} = 0$$

$$M = k\sqrt{L_1L_2}, \quad k = 1; \qquad L_1, L_2 = L \tag{5.6}$$

$$E_d = L\frac{di_L}{dt} + L\frac{di}{dt} + V_L$$

In the next step, keeping in mind symmetry in operation and commutation for each conducting period and taking the Laplace transform of equation 5.6, the following equation is established in the complex frequency domain:[3]

[3] K. Denno, "Steady State and Dynamic Investigations . . .".

$$\frac{I_L(S)}{E_d(S)} = \left(\frac{1}{L}\right)\left(\frac{1}{S} + LCS\right) + C\,\frac{V_C(t_o)}{E_d(S)} - \left(\frac{1}{L}\right)\left(\frac{1}{S} + LCS\right)\frac{V_L(S)}{E_d(S)} + \frac{i_L(t_o) + i(t_o)}{SE_d(S)} \tag{5.7}$$

$$G = \left(\frac{1}{L} + \frac{1}{S}\right) + (SC)$$

$$O = E_{ic} + \frac{1}{C}\int i_c dt + L\,\frac{di_c}{dt} \tag{5.8}$$

where

$$E_{ic} = \text{initial capacitor voltage}$$
$$I_i = \text{initial inductor current}$$

Equation 5.8 can determine the final values of capacitor voltage and current that will be used as initial values for the second commutating interval, when the following relation is valid:

$$E_d = E_c(t_o) + \frac{1}{C}\int i_c dt + L\,\frac{di_c}{dt} \tag{5.9}$$

Taking the Laplace transform of equation 5.9 and considering the first commutating interval effect, the following equation is obtained (symmetry is assumed in the operation and commutation for the total conducting period):[4]

$$\frac{I_L(S)}{E_d(S)} = \frac{1}{2}G - \frac{V_L(S)}{E_d(S)}G - \frac{E_c(t_1)}{E_d(S)}\frac{G}{S} + \frac{I_L(t_1)}{E_d(S)}LG \tag{5.10}$$

where

$$G = \frac{G_1}{1 + G_1H_1} = \frac{1/R[1 + (L/C)S]}{1 + 1/SRC[1 + (L/C)S]} \tag{5.11}$$

$$G_1 = \frac{1}{R[1 + (L/C)S]}$$

$$H_1 = \frac{1}{SC} \tag{5.12}$$

[4] Based on paper presented at the 1976 IEEE MIDWEST POWER SYMPOSIUM. October 1976, Kansas State University. Paper entitled: ''Dynamic Modeling of Basic Types of Solid State Power Inverters'' 37 pages, by K. Denno.

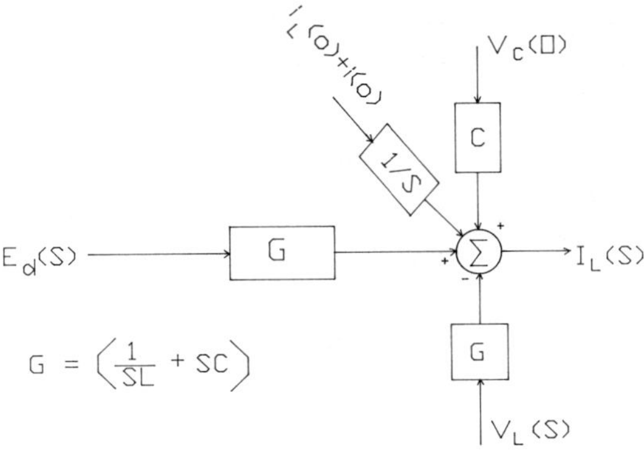

Figure 5.11 Dynamic model of the transformer commutated inverter. (© 1968 IEEE. Humphrey, "Inverter Commutation Circuits.)

And, finally, from the set of equations 5.12–5.17, a dynamic model for the transformer commutated inverter is secured in Fig. 5.11.

An improved design of the transformer commutated inverter developed by J. P. Gibson is characterized by reduced dependency of commutation ability on the DC supply voltage and a lower order in the time rate of change in current through the thyristors. Figure 5.12 shows the improved transformer commutated inverter with components that provide commutation ability of 380 amps at a minimum DC voltage of 62 volts. A commutation transformer ratio of 4 is used to regulate the low-current auxiliary power supply. The *di*/*dt* in the commutation process is 7 amp/sec.

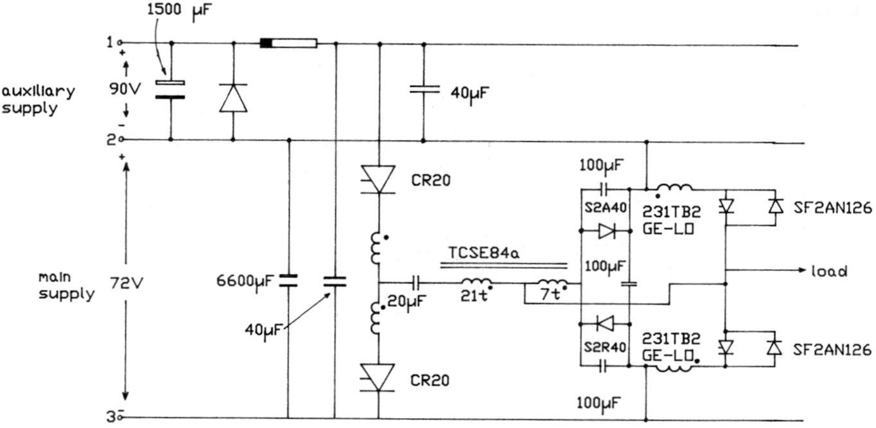

Figure 5.12 Circuit diagram of improved experimental inverter. (© 1976 IEE. Reprinted with permission from J. P. Gibson, "New Inverter Circuit Suitable for High Current P. W. M. Operation.")

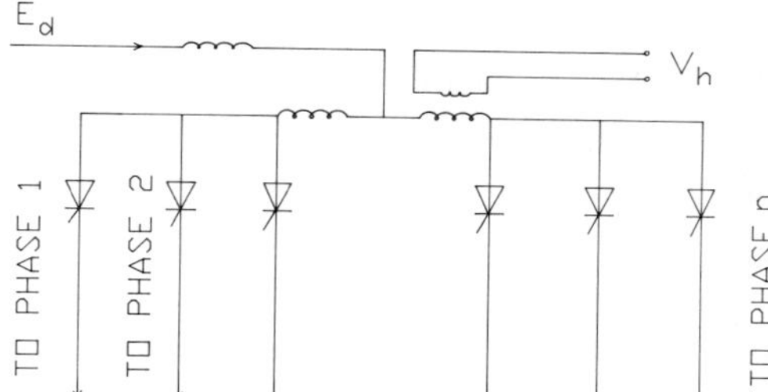

Figure 5.13 An *n*-phase externally harmonic commutated inverter. (Reprinted with permission: © 1934 IEEE, C. H. Willis and Bedford & Holt (John Wiley & Sons, Inc., 1964).

5.3.4 Modeling of the *n*-Phase Externally Harmonic Commutated Inverter

Circuit Diagram Operation and Modeling

Because direct-line commutation cannot supply an adequate inductive KVA load, harmonic commutation, with phase-control angles ranging from 180° to 270°, can place a reliable loading power factor on the AC lines. This is especially effective with static loads.

The *n*-phase harmonic commutated inverter shown in Fig. 5.13 has the following features:[5]

1. Ideal transformers, reactors, and SCRs.
2. Large DC inductance to sustain continuous DC current.
3. AC line voltage independent of the inverter.
4. An adjustable DC supply voltage that may change with the load.
5. A pure sinusoidal harmonic voltage source with amplitude and phase of the harmonic voltage corresponding to the firing angle of the SCR. When commutation occurs near the maximum value of the harmonic voltage wave, the external harmonic voltage will be at a minimum.

Figure 5.14 illustrates one stage in the operation of this type of commutated inverter.

Dynamic Model

Dynamic modeling is based on consideration of the commutation process and the fact that the current in each SCR must vanish prior to the end of

[5] From C. H. Wills, "Applications of Harmonic Commutation for Thyration Inverters and Rectifiers," *AIEE Transactions,* 1934, pp. 701–706 and B. D. Bedford and R. C. Holt, *Principles of Inverter Circuits* (New York, N.Y.: John Wiley & Sons, Inc., 1964).

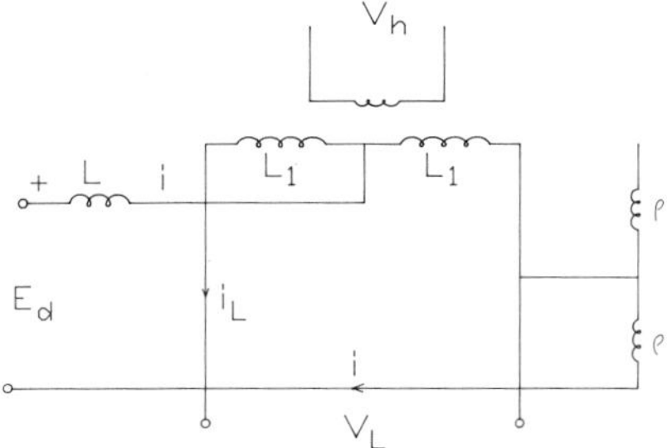

Figure 5.14 Single state of performance for the *n*-phase externally harmonic commutated inverter. (Reprinted with permission: © 1934 IEEE, C. H. Willis and Bedford & Holt (John Wiley & Sons, Inc., 1964).

each conducting interval. With reference to Fig. 5.14 we can write the following circuit equations in the time domain:

$$E_d = (L + L_1 + l)\frac{di}{dt} - L_1\frac{di_L}{dt} + \frac{1}{2}V_h \tag{5.13}$$

$$V_L = -L_1\frac{di}{dt} + 2L_1\frac{di_L}{dt} - V_h \tag{5.14}$$

Assuming that the external harmonic voltage frequency is of the second harmonic order, we can write

$$V_h(t) = V_{hm}\sin 2\omega t \tag{5.15}$$

By taking the Laplace transforms of the above-mentioned equations and recognizing the principle of commutation, the following dynamic equation is obtained:

$$\begin{aligned}\frac{I_L(S)}{E_d(S)} &= \frac{k_1}{s} + \frac{V_L(S)}{E_d(S)}\frac{k_2}{s} - \frac{V_h(S)}{E_d(S)}k_3S + \frac{k_3}{E_d(S)}V_{ho} \\ &+ \frac{k_3}{SE_d(S)}V_{ho} + \frac{i_{Lo}}{E_d(S)}\frac{1}{S}\end{aligned} \tag{5.16}$$

where

i_{Lo} = initial value of the pulsing current
V_{ho} = initial value of the harmonic voltage
V_{ho} = initial value of the first-time derivative for the harmonic voltage

$$k_1 = \frac{l + L_1}{L(l + L)} \tag{5.17}$$

$$k_2 = \frac{\rho + L + L_1}{L(l + L)} \tag{5.18}$$

$$k_3 = \frac{CL_1}{l + L} \tag{5.19}$$

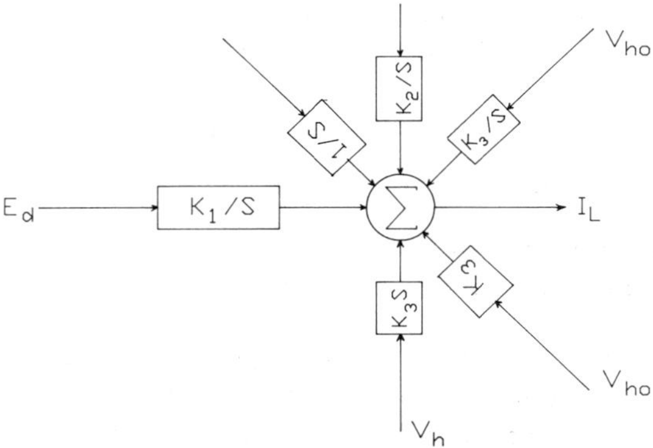

Figure 5.15 Dynamic model of the *n*-phase externally harmonic commutated inverter. (Reprinted with permission of Energy Utilization of P.S.E. & G. of NJ; 1976 IEEE MIDWEST POWER SYMPOSIUM.)

Now by virtue of equation 5.16, we can move to represent the dynamic model of this inverter in the (S) domain, as shown in Fig. 5.15.

5.3.5 Modeling of the n-phase Internally Commutated Inverter

Circuit Diagram and Operation

A harmonic voltage can be provided by capacitive impedance. Small amplitude can establish commutation when operated in the vicinity of phase position for AC commutation. However, considerable harmonic voltage is required for operation beyond the 0–180° interval of phase-control angles. The capacitive impedance must be large enough for any load current to supply the required harmonic voltage.

Figure 5.16 illustrates a double-way version of a six-phase harmonic commutation, which is preferable for applications connected to high-voltage

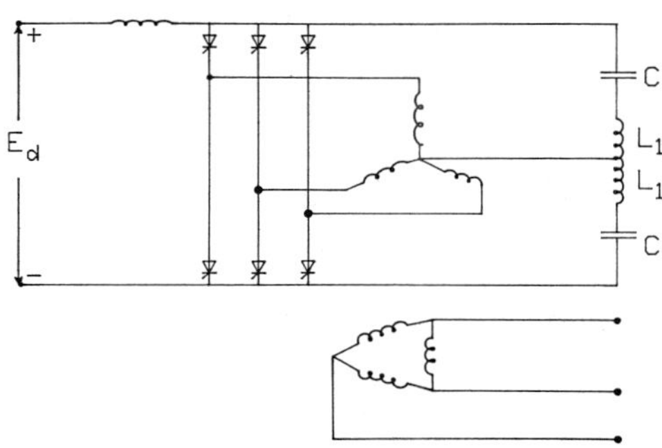

Figure 5.16 Circuit diagram of the *n*-phase harmonic internally commutated inverter. (© 1964 John Wiley & Sons, Inc. Reprinted, with permission from the book: Principles of Inverter Circuits, by B. D. Bedford and R. C. Holt, 1964.)

DC voltage. Figure 5.17 illustrates one stage of operation in which SCR_1 conducts over a 60° interval in contrast to the 120° period in the conventional types.

Dynamic Model

As we stated before, modeling is always based on the assumption that the current in each SCR must vanish prior to the end of each conducting interval.

Now we can write the basic equations of operation of the n-phase harmonic internally commutated inverter using the circuit in Fig. 5.17 and keeping track of the commutation process as expressed below:

$$E_d = L \frac{di}{dt} + V_L \tag{5.20}$$

$$\begin{aligned} V_h + V_L = &\frac{1}{c} \int (i - i_L - i_s)\, dt + L_1 \frac{d}{dt} (i - i_L - i_s) \\ &+ \frac{1}{c} \int (i - i_L)\, dt + L_1 \frac{d}{dt} (i - i_L) \end{aligned} \tag{5.21}$$

$$\frac{1}{2} (V_h + V_L) = \rho \frac{di_s}{dt} \tag{5.22}$$

And assuming that the internally generated harmonic voltage is of third harmonic frequency order,

$$V_i(t) = V_{Lm} \sin 3\omega t \tag{5.23}$$

In the next step, taking the Laplace transform of equations 5.20–5.23, a mathematical model in the (S) domain is secured as in equation 5.24, from which a block diagram representing the dynamic control model is established as shown in Fig. 5.18.

$$\begin{aligned} \frac{I_L(S)}{E_d(S)} = &\frac{1/L}{S(1 + L_1 CS)} - \frac{V_L(S)}{E_d(S)} G_1 + \frac{V_h(S)}{E_d(S)} G_2 \\ &+ \frac{M_2 L}{E_d(S)} G_3 - \frac{M_3 C}{E_d(S)} G_4 - \frac{M_1 C}{E_d(S)} G_5 + \frac{U_{Lo}}{E_d(S)} G_6 \\ &- \frac{U_{ho}}{E_d(S)} G_7 - \frac{i_{Lo}}{E_d(S)} G_8 \end{aligned} \tag{5.24}$$

where

$$G_1 = \frac{Ck_2[1 + (k_1/k_2)S^2]}{S(1 + SL_1C)} \tag{5.25}$$

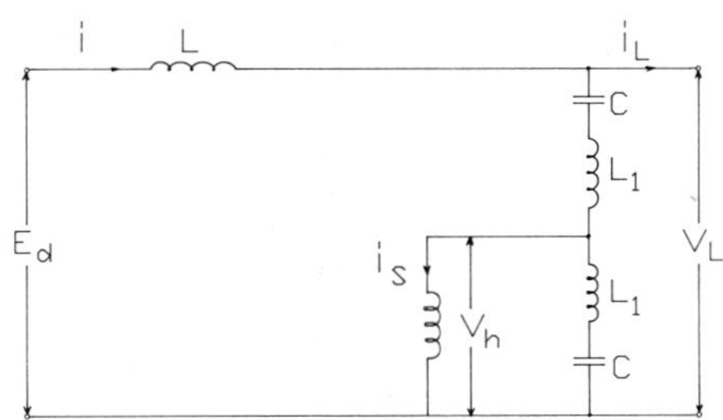

Figure 5.17 Single-stage circuit representation of six-phase harmonic internally commutated inverter. © 1964 John Wiley & Sons, Inc. Reprinted, with permission from the book: Principles of Inverter Circuits, by B. D. Bedford and R. C. Holt, 1964.

$$G_2 = \frac{Ck_4[1 + (k_3/k_4)S^2]}{S(1 + L_iCS)} \tag{5.26}$$

$$G_3 = \frac{1}{S(1 + L_1CS)} \tag{5.27}$$

$$G_4 = \frac{k_4}{S(1 + L_1CS)} \tag{5.28}$$

$$G_5 = \frac{k_2}{S(1 + L_1CS)} \tag{5.29}$$

$$G_6 = \frac{Ck_1}{(1 + L_1CS)} \tag{5.30}$$

$$G_7 = \frac{Ck_3}{(1 + L_1CS)} \tag{5.31}$$

$$G_8 = \frac{CL_1}{(1 + SL_1C)} \tag{5.32}$$

$$k_1 = 1 + \frac{L_1(2 + 2)}{2L} \tag{5.33}$$

Application innovation in the effective utilization of the internally n-phase harmonically commutated inverter is the split-phase variable fre-

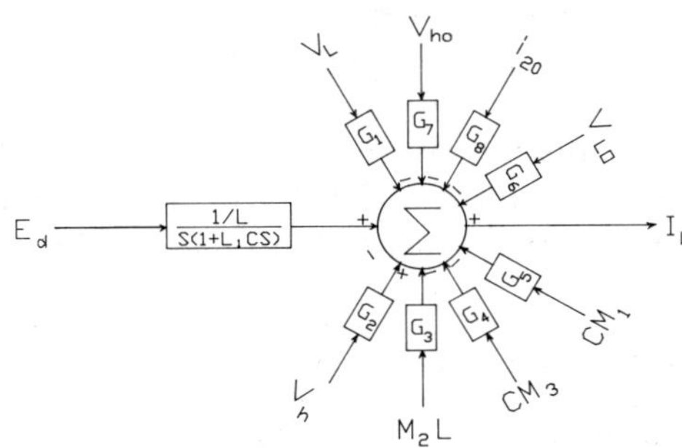

Figure 5.18 Dynamic model of the n-phase harmonic internally commutated inverter. (Reprinted with permission of Energy Utilization of P. S. E. & G. of NJ; 1976 IEEE MIDWEST POWER SYMPOSIUM.)

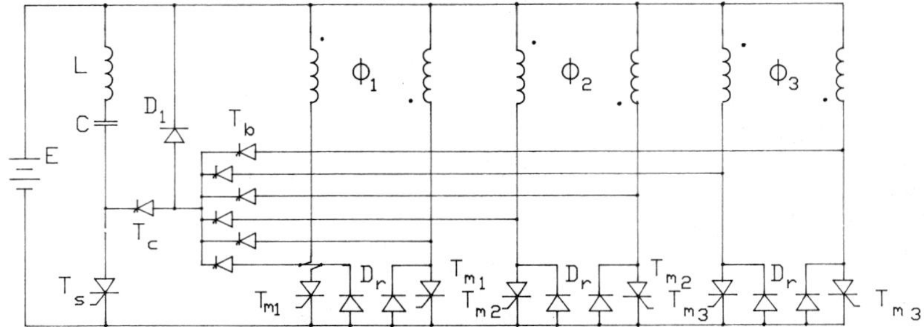

Figure 5.19 Split-phase variable frequency inverter. (© 1982 IEE. Reprinted, with permission, from IEE Proceedings, Vol. 129, pt. B, No. 6, pp. 353–54, November 1982. Paper entitled: "Variable Frequency Inverter" by B. W. Williams.)

quency inverter developed by Williams as shown in Fig. 5.19, which is intended to drive three-phase induction machines.

$$k_2 = \frac{2 + L}{2LC} \tag{5.34}$$

$$k_3 = \frac{2 + L_1}{2}, \quad k_4 = \frac{1}{2C} \tag{5.35}$$

5.4 SUMMARY[6]

In this chapter the physical and operational concepts of solid state power inverters of almost all modes of commutation have been discussed. The principle of the process of commutation for each mode of inversion has been presented with conclusive indication of relative superiority of one mode of commutation with respect to others.

Since the power inverter will become an integral operational stage of an alternative source of power system for an ac output, either performing independently or through an interconnection with other static power systems or the conventional which is electromechanical, modeling of all types of power inverters in the complex frequency domain has been presented in this chapter.

Mathematical model of every solid state inverter serves as direct tool to secure information for its steady-state and time response subject to the application of any kind of electrical or mechanical forcing functions.

[6] © 1974 Middle Atlantic Power Research Committee. Ideas and Reprints are used from the Technical Report of June 1974. Report entitled: "Steady State and Dynamic Investigations for Determining Optimum Electrochemical—Electromechanical Interconnected Power System" by K. Denno.

5.5 SOLVED EXAMPLES

1. Fig. 5.8 represents dynamic model for parallel the capacitor commutated inverter. Under the condition $L_1 + 2L = 0$, solve for $i_L(t)$.

Solution Rewrite equation 5.5 for $L_1 + 2L = 0$.

$$\frac{I_L(S)}{E_d(S)} = \frac{1}{2LS} + \frac{I_1(0)}{E_d(S)}\frac{1}{S} \tag{1a}$$

since,

$$E_d(S) = \frac{E_d}{S} \tag{1b}$$

Therefore (1a) becomes,

$$I_L(S) = \frac{E_d}{2L}\frac{1}{S^2} + \frac{I_1(0)}{S} \tag{1c}$$

hence,

$$i_L(t) = \frac{E_d}{2L}\, t + I_1(0)U_{-1}(t) \tag{1d}$$

From (1d), $i_L(t)$ is the superposition of ramp and a step functions.

2. Fig. 5.11 represents dynamic model for the transformer commutated inverter, solve for $i_L(t)$ under accidental short-circuit at the load. Assume $G = k(R,L,C)S$.

Solution Rewrite equation 5.10, where $V_L(S) = 0$ and $G = kS$. Therefore,

$$I_L(S) = \frac{E_d(S)}{2}\, kS - kE_c(t) + I_L(t_1)kLS \tag{2a}$$

hence, since $E_d(S) = \dfrac{E_d}{S}$, (2a) becomes

$$I_L(S) = \frac{kE_d}{S} - kE_c(t_1) + kI_L(t_1)LS \tag{2b}$$

Therefore,

$$i_L(t) = \left[\frac{kE_d}{2} - kE_c(t_1)\right] U_0(t) + kL\, I_L(t_1)U_0'(t) \tag{2c}$$

We see that the short-circuiting current in the time domain is the superposition of impulse and douplet functions.

3. Repeat example 2, but for an open-circuit load solve for $v_L(t)$.

Solution Review equation 5.10, where $I_L(S) = 0$.

$$\frac{V_L(S)g}{E_d(S)} = \frac{g}{2} - \frac{E_c(t_1)}{E_d(S)}\frac{g}{S} + \frac{I_L(t_1)}{E_d(s)}\, Lg \tag{3a}$$

or

$$V_L(S) = \frac{1}{2} E_d(S) - \frac{E_c(t_1)}{s} + I_L(t_1)L \tag{3b}$$

Since $E_d(S) = \frac{E_d}{S}$, (3b) becomes,

$$V_L(S) = \frac{E_d}{2S} + \frac{E_c(t_1)}{S} + I_L(t_1)L \tag{3c}$$

and

$$v_L(t) = \left[E_c(t_1) + \frac{E_d}{2}\right] U_{-1}(t_1) + LI_L(t_1)U_0(t) \tag{3d}$$

where,
$U_0(t)$ and $U_{-1}(t)$, an impulse and step function respectively.

4. Fig. 5.15 represents dynamic model for n-phase externally commutated inverter. For $V_{ho} = 0$ and under short-circuit load, solve for $i_L(t)$.

Solution Rewrite equation 5.16 for $V_L(S) = V_{ho} = 0$

$$\frac{I_L(S)}{E_d(S)} = \frac{k_1}{S} - \frac{V_h(S)}{E_d(S)} k_3S + \frac{i_{Lo}}{E_d(S)} \frac{1}{S} \tag{4a}$$

or

$$I_L(S) = \frac{k_1E_d(S)}{S} - k_3V_h(S)S + \frac{i_{Lo}}{S} \tag{4b}$$

Therefore,

Since $E_d(S) = \frac{E_d}{S}$, $V_h(S) = \frac{2w}{S^2 + 4\omega^2}$

(4a) becomes,

$$I_L(S) = \frac{k_1E_d}{S^2} - k_3V_{hm} \frac{2\omega S}{S^2 + 4\omega^2} + \frac{i_{Lo}}{S} \tag{4c}$$

Solution of load short-circuiting current shows superposition of a step, ramp and sinusoidal functions.

5. Another well-known model of inverter is the auxiliary-impulse commutating type in which the commutating pulse current adjusts itself automatically with the load. Figure E.5 represents a simple circuit model for the first half of the commutating process in which SCR_1 is conducting from the upper half of the DC source where capacitance C is already charged to V_{co} from previous commutation. The first cycle of commutation starts with the firing of SCR_{1a} where pulsing current will flow through L and C and in rectifier D_1. Excess current flow in D beyond load level will rise up to peak level after which it starts to decay, generating reverse voltage across C resulting in the turn off of SCR_1.

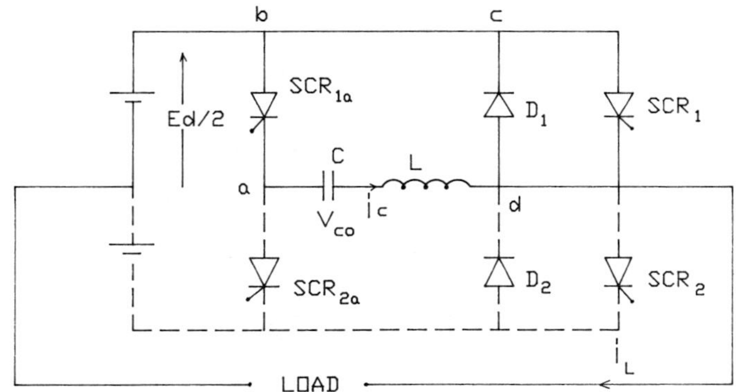

Figure E.5 Circuit diagram for the auxiliary-impulse commutated inverter. (© 1984 John Wiley & Sons, Inc. Reprinted, with permission from B. D. Bedford and R. C. Holt, *Principles of Inverter Circuits*.)

Write the relevant commutating equation of the first cycle with its initial conditions. Also predict writing the commutating equation of the second cycle.

Solution From loop *abcd* for the first commutating cycle, we can write

$$V_{co} + \frac{1}{C}\int i_c\, dt + L\frac{di_1}{dt} = 0 \tag{5.a}$$

where

V_{co} = the initial charged voltage across C
I_{Lo} = the initial inductor current

Consider the second commutating cycle, which commences with the firing of SCR_2 when the pulsing current has been decayed to its minimum value. During this cycle, V_{co} is recharged across capacitance C. This cycle ends with SCR_{1a} having a reversed voltage ceasing to conduct and thus SCR_{2a} is fired. Therefore, the following circuit model equation represents the second cycle of commutation:

$$E_d = V_c(t_o) + \frac{1}{C}\int i_c\, dt + L\frac{di_L}{dt} \tag{5.b}$$

where $V_c(t_o)$ is the newly charged voltage across C.

6. Equation 5.2 represents the mathematical model of the direct-line commutated inverter in the (s) domain. Solve for $i_2(t)$, if $i_1(t_o) = i_2(t_o)$ and $Z_L(s) = SL$.

Solution With $i_1(t_o) = i_2(T_o)$, equation 5.2 becomes

$$I_L(S) = \frac{E_d(S)}{SL} - V_L(S)\frac{L_2}{S} \tag{6.a}$$

$$V_L(S) = I_L(S)SL, \quad E_d(S) = \frac{E_o}{S} \tag{6.b}$$

Therefore,

$$I_L(S) = \frac{E_o}{S^2L} - I_2(S)SL\frac{L_2}{S}$$

$$= \frac{E_o}{S^2L} - LL_2I_2(S) \tag{6.c}$$

$$I_L[(S)1 + LL_2] = \frac{E_o}{S^2L} \tag{6.d}$$

$$I_L(S) = \frac{e_o}{L(1 + LL_2)S^2}$$

Therefore,

$$i(t) = \frac{E_o t}{L(LL_2 + 1)} \tag{6.e}$$

5.6 REVIEW QUESTIONS

1. Write a set of circuit parameters equations in the time domain to represent Fig. 5.1 without feedback control component. Assume that the inverter output = V_m sin ωt.
2. Repeat Question 1 with $V(t) = V_{dc} + V_m \sin \omega t$.
3. Modify equation 5.2 if $0 = \pi/3$ radians.
4. Explain the new inverter circuit that will emerge if the delta inversion is replaced by a star system.
5. Show the implication of an extremely small C for the parallel-capacitor commutated inverter through the set of equations 5.4.
6. For the transformer commutated inverter shown in Fig. 5.10, show the effect on the set of circuit equations 5.6 if the capacitance is very large.
7. Explain the difference between transformer commutated inverter and the parallel-capacitor type.
8. Refer to Fig. 5.12, which is a new design of transformer commutated inverter. Predict the effect if di/dt is about 16 amp/sec instead of 8 amp/sec.
9. Explain the difference between the internally and externally commutated inverter.
10. Refer to Fig. 5.19 showing the split-phase variable frequency inverter. What is the effect on the process of commutation at the resonant frequency?

5.7 PROBLEMS

1. Inversion is considered the reverse process of rectification, with the ultimate aim of securing sinusoidal output voltage or current at the fundamental frequency. Using Fourier series expansion, establish the sinusoidal expansion of the follow-

ing two functions:

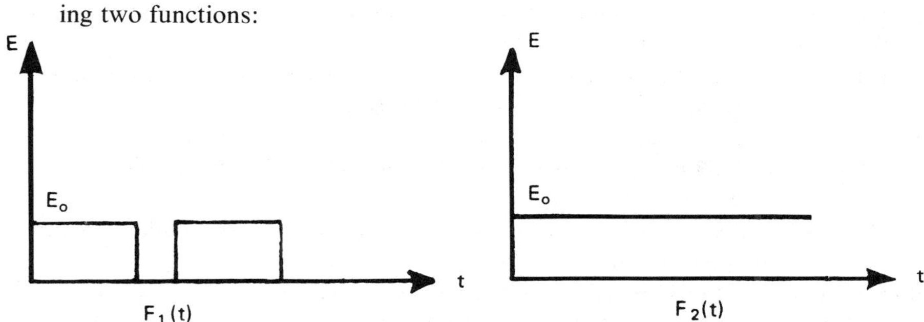

2. a. Expand the train of sinusoids given in Fig. P.1 into a set of harmonics and their fundamentals.
   b. Find the root mean square and the average of this sinusoidal train.
3. Refer to Fig. 5.1, which shows general block diagram for securing pure sinusoidal output from a DC alternative energy device. Establish the mathematical model in the (s) domain under the following conditions:

   inverter output current $= I_{dc} + I_m \sin \omega t$

   ideal transformer of step-up ratio of N

   control feedback function $= kt$

4. For the direct-line commutated inverter shown in Fig. 5.2, carry out systematic steps for the derivation and structuring its model in the complex frequency domain, which is shown in Fig. 5.3.
5. Figure 5.5 shows the newly developed direct-line delta inverter. Using its mathematical model in the (s) domain, develop the three-phase voltage model of the delta inverter in the (s) domain.

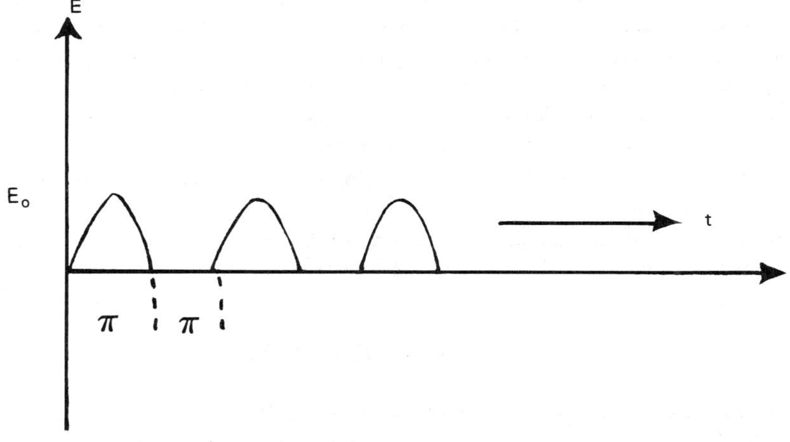

Figure P.1 Train of sinusoids.

6. Develop a unified mathematical model in the (s) domain for a fuel cell coupled to the direct-line commutated inverter.

7. Develop a unified mathematical model in the (s) domain for the storage battery coupled to the direct-line commutated inverter.

8. Use the relevant circuit equations of the parallel-capacitor inverter and Laplace's transform to establish its mathematical model in the (s) domain expressed in equation 5.5.

9. Develop a unified mathematical model in the (s) domain for the parallel capacitor commutated inverter coupled to the fuel cell.

10. Develop a unified mathematical model in the (s) domain for the parallel-capacitor commutated inverter coupled to the storage battery.

11. Carry out systematic steps of Laplace transformation on the set of circuit equations 5.6 for the transformer commutated inverter to establish its mathematical model in the (s) domain expressed in equation 5.7.

12. Establish a unified model in the (s) domain for the interconnection of the transformer commutated inverter to the conventional fuel cell.

13. Establish a unified model in the (s) domain for the interconnection of the transformer commutated inverter with the storage battery.

14. Show systematic steps of Laplace transformations to establish the mathematical model of the n-phase externally harmonic inverter as shown in equation 5.16.

15. Establish a unified mathematical model in the (s) domain for the interconnection of the n-phase externally commutated inverter with the conventional fuel cell.

16. Carry out systematic steps of a Laplace transformation to secure the mathematical model in the (s) domain for the n-phase internally commutated inverter.

17. Establish a unified mathematical model for the interconnection of the n-phase internally commutated inverter with the conventional fuel cell.

18. Establish a unified mathematical model in the (s) domain for the interconnection of the n-phase internally commutated inverter with the storage battery.

19. Develop a unified mathematical model in the (s) domain for the interconnection of the series-capacitor commutated inverter with the conventional fuel cell.

20. Draw a circuit diagram for the series capacitor commutated inverter where the inductor is divided into two equal parts. Then write the circuit commutation equations in the time domain and in the frequency domain with zero initial conditions. Develop a working transform function.

21. Repeat problem 20 for the series capacitor inverter where the capacitance is divided equally.

22. Develop a unified mathematical model in the (s) domain for the series-capacitor commutated inverter linkage to the storage battery.

5.8 REFERENCES

1. A Power Special Report, *Power Journal,* March 1967.
2. Bedford, B. D., and R. G. Holt. *Principles of Inverter Circuits*. New York: John Wiley & Sons, 1964.

3. Bowens, S. R., J. Clare, and R. R. Clements. "Transient Performance of Inverter Systems." *IEE Proceedings* (England), Vol. 129, Pt. B, no. 6 (November 1982), pp. 301–313.
4. Bradley, D. A., C. D. Clarke, and R. M. Davis. "Adjustable Frequency Inverters and Their Application to Variable-Speed Drives." *Proceedings of IEE* (England), Vol. 111 (November 1964).
5. Cornick, J. A., and M. J. Ramsbottom. "Instantaneous Temperature Rise in Thyristors Under Inverter and Chopper Operating Conditions." *Proceedings of IEE* (England), Vol. 119 (August 1972).
6. Evans, P. D., R. C. Dodson, and J. F. Eastham. "Delta Inverter." *IEE Proceedings,* (England), Vol. 127, Pt. B, no. 6 (November 1980), pp. 333–340.
7. Humphrey, A. J. "Inverter Commutation Circuits." *IEEE Transaction on Industry and General Applications,* Vol. 4 (January–February 1968).
8. Kernick, A., J. L. Roof, and T. M. Heinrich. "Static Inverter with Neutralization of Harmonics." *Proceedings of IEE* (England), May 1982.
9. MacDonald, I. M. "A Static Inverter, Wide Range, Adjustable Speed Drive." *IEEE Transactions on Industry and General Applications,* January–February 1960.
10. Mapham, N. W. "The Classification of SCR Inverter Circuits." *IEEE Transactions on Industry and General Applications,* January–February 1960.
11. McMurray, W., and D. P. Shattuch. "A Silicon-Controlled Rectifier Inverter with Improved Commutation." *AIEE,* November 1961.
12. Subrahmanyam, V. "Analysis of Commutation of a Current Source Inverter Feeding an Induction Motor Using 2-Axis Variables." *IEE Proceedings* (England), Vol. 129, Pt. B, no. 6 (November 1982), pp. 353–354.
13. Ward, E. E. "Inverter Suitable for Operation over a Range of Frequency." *IEE Proceedings* (England), Vol. 111 (August 1964).
14. Williams, B. W. "Variable-Frequency Inverter." *IEE Proceedings* (England), Vol. 129, Pt. B, no. 6 (November 1982), pp. 353–354.

6

Ocean Thermal Energy Conversion (OTEC)

6.1 INTRODUCTION

Efficient utilization of ocean thermal energy conversion (OTEC) plants depends centrally on the development of an effective and economical system to transport the OTEC output to the locations of power demands. The OTEC system is based on the volumetric availability of bulk ocean temperature gradient at the ocean plant floating location or in close proximity such as on a land base resulting in the production of hydrogen and oxygen through endothermic decomposition of water and the release of nitrogen through air liquefaction. The multioutputs of the OTEC plant—namely, ΔT, hydrogen, oxygen, and nitrogen—are used to generate mechanical power to drive conventional electric power plants, or the OTEC plant uses hydrogen as the main fuel for fuel-cell systems and may utilize the huge bulk of ocean ΔT to preheat the working fluid in magnetohydrodynamic power plant. The preferred effective value of ΔT in the tropics is between the warm surface temperature, on the order of 25°C, and the cold bottom water 500–1000 meters below, at around 4°C. Usually the 20–21°C temperature gradient is distributed through an oceanic area around 10 km off-shore, and hence the OTEC plant is more likely to be a floating station, although where coastal bathymetry permits, OTEC land-built plants could be set up. Maximum theoretical efficiency of an OTEC plant is about 7.5 percent, and the actual operational efficiency is substantially less than that.

Research for the technical and commercial feasibility of an OTEC system is progressing in the United States, Japan, and Taiwan. In the

United States two bills have been signed by President Carter: The Ocean Thermal Energy Conversion Research and Development Act and the Ocean Thermal Conversion Act. The U.S. OTEC research program calls for the initial testing of a 40-MW_e OTEC plant, followed by an 100-MW_e plant in 1986 and 500-MW_e plant in 1989.[1]

Utilization of OTEC output could be in the production of fuel cells such as ammonia or hydrogen and material such as aluminum, or could be in the generation of electric energy to be transported through conventional or indirect means for utilization at any destination.

6.2 OTEC SYSTEMS[2]

OTEC plants could be an open-cycle system, where the warm seawater is evaporated at low pressure, producing low-pressure steam for turbine drive, and the closed-cycle plant, where the ocean bulk ΔT will be used to produce working fluid with boiling temperature ranging between 4° and 20°C at normal pressure, such as ammonia, propane, or any operational refrigerant.

Main components of a closed-cycle OTEC plant are

1. Heat exchanger–condenser unit
2. Heat exchanger–evaporater unit
3. Cold water pipe (CWP) system
4. Electric grid system

For heat exchangers, it is suggested that an effective area of 2.5 km^2 is required for a 400-MW_e plant, while maximum design area at the present state of the art is commonly around 5000 m^2. Research efforts for an effective OTEC plant are primarily targeting the optimization of operational area of the heat exchanger for maximum surface interface and minimum weight.

The most suitable material for heat exchangers is titanium, which offers maximum resistance to corrosion and excellent compatibility to closed-cycle working fluids such as ammonia; however, it is relatively expensive with respect to aluminum, stainless steel, and plastics.

Condenser units act to return the working fluid such as ammonia to its initial liquid state through the circulation of cold water lifted through a large piping system from the bottom of the ocean, which is normally at a temperature of around 4°C.

For the CWP, or the cold water pipe, system, it is suggested that

[1] © 1983 IEE. Reprinted, with permission, from IEE Proceedings, Vol. 130, pt. A, No. 2, pp. 93–100, March 1983. Paper entitled: "Ocean Thermal Energy Conversion" by G. Ford, C. Niblett and L. Walker.

[2] Ibid.

polythene is very suitable for the process of design and construction. However, at the present it is estimated that for a 400-MW_e plant, a CWP system 30 m in diameter and over 900 m length is required, which is far from being physically realized with the available technology. Actually a CWP design set by Global Marine made of polythene is 1 m in diameter and 600 m in length. Hence it can be said that intensive research work is imperative to secure commercial realization of a large CWP system, matching the operational design of a multihundred MW_e plant.

Loads on the CWP have been identified, but their actual introduction in the design process is still under intensive research work. These loads are

1. Current drag
2. Water forces due to particle acceleration
3. Oscillating forces from vortex shedding
4. Forces due to the internal flow in the CWP
5. Forces due to slow drift of OTEC station
6. Dead-weight forces

Turning to the electric grid system, this could be modeled as expressed in the following discussion.

6.2.1 Direct System

Here electric power is produced by an AC generator driven by the gas turbine. The working gas or vapor is the result of converting the working fluid such as ammonia to the vapor state. The produced electric power output could be transmitted through cable and an overhead system to any location for energy demand. Transportation of electric energy could be accomplished via chemical means, for example, by recharging a redox flow cell; transforming its catholyte ferrous solution into ferric solution, storing the electric energy in chemical form; and then shipping the condensed ferric solution to the place where it can be diluted to the required level of concentration and then used in the catholyte of the redox cell during the discharge period to supply power to the electric load.

6.2.2 Indirect System

Here the huge bulk ΔT could be used in these processes:

1. Endothermic decomposition of water into H_2 and O_2
2. Air liquefaction and the extraction of nitrogen
3. Using the bulk ΔT to raise the temperature of the magnetohydrodynamic (MHD) generator working gas, and hence an MHD plant could

be coupled to OTEC for the generation of electric power locally or at any place far from OTEC station

4. Using hydrogen as fuel for the conventional fuel cell either locally or away from OTEC system

6.3 ISENTROPIC THERMAL ENERGY EXTRACTION

Isentropic heat exchange process implies that all changes are subject to zero pressure and temperature gradients within the exchangers.

6.3.1 Current Technology

Today's technology still relies heavily on using conventional heat exchangers (condenser and evaporator) as the major components for the OTEC plant. However, innovation in design of heat exchangers through continuous research is progressing at various centers in Hawaii, Gulf of Mexico, Puerto Rico, Japan, and Taiwan as well as at many research laboratories seeking the development of effectively operational and economically feasible heat exchangers. Work conducted at the Applied Physics Laboratory at Laurel, Maryland, by P. Pandolfini, G. L. Dugger, F. K. Hill, and W. H. Avery and H. D. Faust at the Trane Company of LaCrosse, Wisconsin developed a prototype heat exchanger modeled with Alclad aluminum tubes folded in a set of passes in a horizontal plane configuration. For a core heat exchanger unit compatible for a 30 kW_e plant, the measured overall heat transfer coefficients for a condenser unit was 2900 W/m^2-K and for an evaporator unit 2250 W/m^2-k°, recorded at design conditions.

At the same time, the work at Laurel, Maryland, and LaCrosse, Wisconsin, established a very important analytical and experimental correlation for the heat transfer coefficients with respect to the oceanic Reynolds number, average and variable water velocities, as well as ammonia Reynolds number.[3] These are shown in Figs. 6.1 and 6.2 from which important relationships are as follows:

$$U_{\text{ave}} = 7.30 R_w^{0.4} \tag{6.1}$$

Equation 6.1 is for a condenser heat exchanger unit; equations 6.2 and 6.3 are valid for an evaporator heat exchanger model.

$$U_{\text{pred}} = 0.217\ (R_w R_{NH_3})^{0.2345} \tag{6.2}$$

$$U_{\text{most}} = 0.0743\ (R_w R_{NH_3})^{0.4425} \tag{6.3}$$

[3] "Al Clad—Aluminum Folded—Tube Heat Exchangers for OTEC" by P. Pandolfini, G. L. Dugger, F. K. Hill, W. H. Avery and H. D. Faust in Proceedings of Condensed Papers—3rd. Miami International Conference on Alternative Energy Sources. Dec. 1980.

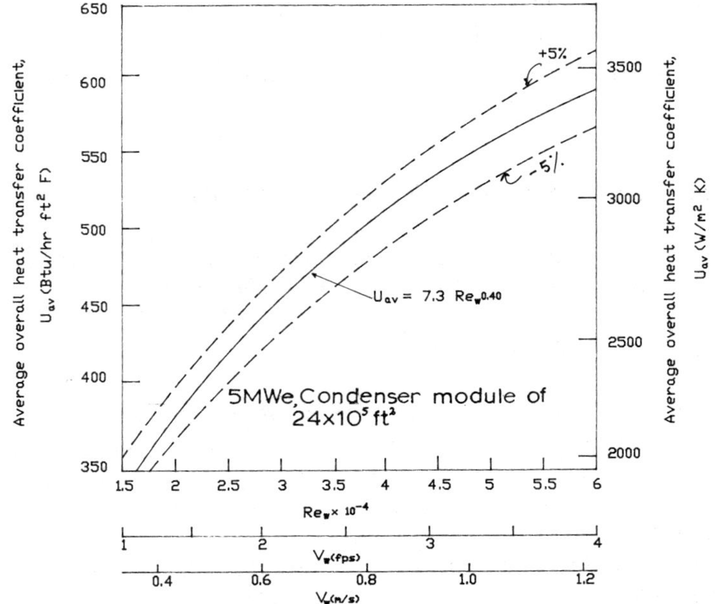

Figure 6.1 Measured overall heat transfer coefficient (U_{av}) versus water flow Reynolds number (R_w) and water velocity in minimum tube gap (v_w) for condenser core test. (from Proceeding of Condensed Papers—3rd Miami International Conference on Alternative Energy Sources: paper by Pandolfini, Dugger, Hill, Avery and Faust.)

where

R_w = the ocean water Reynolds water
R_{NH_3} = the ammonia Reynolds number
U_{pred} = the predicted overall heat transfer coefficient
U_{most} = the maximum or most expected heat transfer coefficient

6.3.2 Water and Ammonia Reynolds Numbers[4]

We can show the ocean water Reynolds number R_w as

$$R_w = \frac{X_w V_w}{\nu_w} \tag{6.4}$$

where

X_w = the reference space variable of water continuum movement
V_w = ocean water state velocity
ν_w = the seawater viscosity

[4] Reprinted, with permission, from Alternative Energy Sources VI, Vol. 3, pp. 141–149, 1985. Paper entitled: "Modelling of Energy Release Systems from OTEC Plants" by K. Denno, © 1985 Hemisphere Publishing Co.

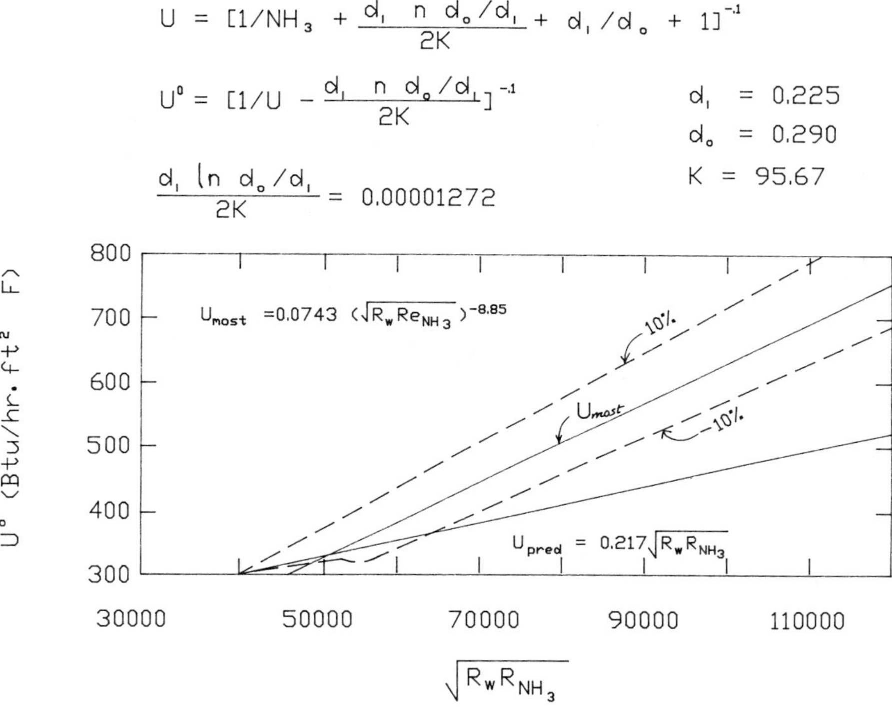

Figure 6.2 Correlation of data for measured overall heat transfer coefficient in evaporator core tests. (from Proceeding of Condensed Papers—3rd Miami International Conference on Alternative Energy Sources: paper by Pandolfini, Dugger, Hill, Avery and Faust.)

Similarly, the ammonia Reynolds number R_{NH_3} is expressed by

$$R_{NH_3} = \frac{X_{NH_3} V_{NH_3}}{\nu_{NH_3}} \tag{6.5}$$

where

X_{NH_3} = the reference space variable of ammonia continuum movement
V_{NH_3} = the ammonia liquid or vapor Reynolds number.

6.3.3 Condenser Unit of Heat Exchanger[5]

Let Q be the rate of heat transfer:

$$Q = UA \frac{(t_1 - t_o) - (t_2 - t_i)}{\ln [(t_1 - t_o)/(t_2 - t_i)]} \tag{6.6}$$

[5] © 1971 McGRAW-HILL BOOK CO. Reprinted, with permission, from the book entitled: DESIGN OF THERMAL SYSTEMS by W. F. Stoecker.

where

U = the overall heat transfer coefficient in Btu/hr ft^2 °F
A = the area of fluid interfacing
t_o, t_i = the outlet and inlet external fluid temperature, respectively
t_2, t_1 = the outlet and inlet internal fluid temperature, respectively

Then from equations 6.1 and 6.6, we can write for the condensate unit,

$$Q_{\text{ave}} = 7.30R_w \frac{(t_1 - t_o) - (t_2 - t_i)A}{\ln [(t_1 - t_o)/(t_2 - t_i)]} \tag{6.7}$$

6.3.4 Evaporative Unit of Heat Exchanger[6]

Rewriting equations 6.2 and 6.3,

$$U_{\text{pred}} = 0.217\ (R_w R_{\text{NH}_3})^{0.2345} \tag{6.8}$$

$$U_{\text{most}} = 0.0743\ (R_w R_{\text{NH}_3})^{0.4425} \tag{6.9}$$

The value of ΔT within the heat exchanger continuum is usually expressed by

$$\Delta T = \left[UY \frac{2T}{MC_p} \right]^{1/2} \tag{6.10}$$

where

U = any heat transfer coefficient
ΔT = extracted total temperature gradient
Y = total time of heat extraction
M = mass of heat carrier in kilograms or pounds
C_p = specific heat at constant pressure

or

$$\Delta T = t_1 - t_o = t_2 - t_i \tag{6.11}$$

Now let

$$\begin{aligned} Q &= \text{total rate of heat transfer} \\ &= AK \end{aligned} \tag{6.12}$$

where

$$K = \frac{(t_1 - t_o) - (t_2 - t_i)}{\ln (t_1 - t_o/t_2 - t_i)} \tag{6.13}$$

[6] Denno, "Magnetic Transport Properties."

Therefore, from the preceding equations,

$$Q_{\text{pred}} = 0.217A(R_w R_{\text{NH}_3})^{0.2345} \left[\frac{(t_1 - t_o) - (t_2 - t_i)}{\ln (t_1 - t_o/t_2 - t_i)} \right] \tag{6.14}$$

and

$$Q_{\text{most}} = 0.0743(R_w R_{\text{NH}_3})^{0.4425} \left[\frac{(t_1 - t_o) - (t_2 - t_i)}{\ln (t_1 - t_o/t_2 - t_i)} \right] \tag{6.15}$$

6.3.5 Solution of R_{NH_3} and R_w

Predicted Case

From equations 6.2, 6.3, and 6.8,

$$R_{\text{NH}_3}^{0.2345} = \left(\frac{9.14YTC_P}{MA^2k^2} \right) R_w^{0.531} \tag{6.16}$$

Let

$$k'' = \frac{9.15YTC_p}{MA^2k^2} \tag{6.17}$$

Upon the application of successive partial self-differentiation with respect to time in equation 6.16, the following is obtained;

$$\frac{\partial R_{\text{NH}_3}}{R_{\text{NH}_3}} = \frac{20}{AK} \frac{0.217}{U_{\text{pred}}} (\partial Q) - \frac{\partial R_w}{R_w} \tag{6.18}$$

From equation 6.18 we can write

$$\frac{\partial R_{\text{NH}_3}}{R_{\text{NH}_3}} - \frac{\partial R_w}{R_w} = \frac{20}{AK} \frac{0.217}{U_{\text{pred}}} \partial Q \tag{6.19}$$

The observation from equation 6.19 is that any difference in normalized perturbation for R_{NH_3} and R_w is affected inversely by changes in the predicted heat transfer coefficient, differences in the internal and external fluid temperatures, and effective interfacing area of heat exchange and is influenced directly by a perturbation in thermal energy released or absorbed.

Equation 6.19 could be used also to express normalized perturbation in ammonia velocity as well as water velocity (variable of average):[7]

$$\frac{\partial V_{\text{NH}_3\text{-ave}}}{V_{\text{NH}_3\text{-ave}}} - \frac{\partial V_{w\text{-ave}}}{V_{w\text{-ave}}} = \frac{20}{AK} \frac{0.217}{U_{\text{pred}}} \tag{6.20}$$

[7] © 1985 Hemisphere Publishing Co. K. Denno, "Modeling of Energy Release Systems from OTEC Plants."

where from equation 6.2

$$U_{\text{pred}} = 0.217(R_w R_{\text{NH}_3})^{0.2345}$$

U_{pred} is the predicted heat transfer coefficient and

$$\partial Q_{\text{pred}} = 0.1465Akk'R_{\text{NH}_3}^{-0.3245}\partial R_{\text{NH}_3} \tag{6.21}$$

where ∂Q is the perturbation in the predicted total heat exchange. Therefore, from equations 6.18 and 6.21, another link between R_w and R_{NH_3} is established:

$$R_{\text{NH}_3} = R_w^{-1.2892k'k''} \tag{6.22}$$

where

$$k' = \left(\frac{MA^2k^2}{9.15YTC_p}\right)^{1.833} \tag{6.23}$$

Linking equations 6.16 and 6.22, the ultimate solution for R_{NH_3} is

$$R_{\text{NH}_3} = k'' \frac{5.55(Ak)^{2.90}(M/9.15YTC_p)^{1.45}}{1 + 1.31(Ak)^{2.90}(M/9.15YTC_p)} \tag{6.24}$$

Corresponding associate solution could be obtained for R_w from equation 6.16, where[8]

$$R_w = \left(\frac{MA^2k^2}{9.15YTC_p}\right)^{1.833} [(R_{\text{NH}_3})^{0.441}] \tag{6.25}$$

Most Probable Case

$$U_{\text{most}} = 0.0743(R_w R_{\text{NH}_3})^{0.4425}$$

and

$$Q_{\text{most}} = U_{\text{most}}Ak \tag{6.26}$$

Procedures similar to the predicted case could be followed in establishing solutions for R_{NH_3} and R_w for the most expected case, which is left as a problem for the student to solve.

6.4 LINKAGE FOR NITROGEN AND HYDROGEN GENERATION[9]

Let ΔG be the perturbation in the Gibbs free energy function or maximum energy release at standard pressure and temperature, which can be linked to

[8] Ibid.

[9] Pandolfini, Dugger and Hill, "Al-Clad Aluminum . . .".

either a predicted or most (optimum) case. Therefore,

$$\Delta G \propto Q_{\text{pred}} \quad \text{or} \quad \propto Q_{\text{most}} \tag{6.27}$$

where $\propto$ means proportional to.

We can write for the predicted case,

$$\Delta G \propto (R_w R_{\text{NH}_3})^{0.2345}$$

Also

$$\Delta G \propto \ln \frac{P_N^a P_H^b}{P_{\text{NH}_3}^c} \tag{6.28}$$

where P_{N}, P_{H}, and P_{NH_3} are the activity or partial pressure for nitrogen, hydrogen, and ammonia, respectively. Also $\propto$ indicates proportional to and a, b, and c represent the molecular or atomic concentration for nitrogen, hydrogen, and ammonia, respectively.

The criterion or controlling function for the mutual release of N, H_2, and NH_3 is now

$$P_{\text{NH}_3}^c[\exp(R_w R_{\text{NH}_3})^{0.2345}] \propto R_{\text{N}}^a P_{H_2}^b \tag{6.29}$$

We have to remember that equation 6.29 is valid for OTEC plant operating under isentropic-evaporative conditions.

The criterion stated in equation 6.29 could be rewritten also in the following form:

$$\frac{P_{\text{N}}^a P_{\text{H}_2}^b}{P_{\text{NH}_3}^c} = \gamma \exp(R_w R_{\text{NH}_3})^{0.2345} \tag{6.30}$$

where γ is a system constant.

6.5 THE FOAM OTEC SYSTEM[10]

Research conducted at Carnegie-Mellon University and at the University of Puerto Rico by Zenner and Kay is targeting the development of a new innovative kind of OTEC system, known as the foam model. This model is based on the elimination of the conventional heat exchanger and the replacement of the low-density steam-turbine with the standard hydroturbine. The foam OTEC system is simply a two-phase vertical-column flow, driven by the series coupling of the total pressure gradient in conjunction with the temperature and density gradients. The tightly coupled two-phase system is water–water vapor, where the water content of the foam will drive the turbine.

[10] Ibid.

Expended work per unit mass expressed approximately by the formula

$$W \approx \frac{p(\Delta T)^2}{2T} \tag{6.31}$$

Equation 6.31 is linked to the condition $\Delta S = 0$, where C_p is the specific heat at constant pressure and ΔS is the entropy perturbation. Experimental research data indicated that energy extraction at 20°C ΔT is 2.90 J/gm of warm water, equivalent to raising ocean water to about a 200-m head.

6.6 SUMMARY

In this chapter general identification for the OTEC open-cycle and closed-cycle systems were presented. Problems of design and operation for bulk OTEC plants concern the effective heat exchanger and the cold water pipe systems for the closed-cycle plant. Productive and operational feasibility for a floating or land-based OTEC system is considered mostly hinging on maximizing the performance of the heat exchanger, the condenser, and the evaporator units. Relatively broad analytical models for the operational characteristics of the heat exchanger were reviewed, for example, the ocean water and ammonia Reynolds numbers, namely, R_w and R_{NH_3}, respectively. Each Reynolds number is a function of the fluid average or variable velocity, the respective reference space dimension, and fluid viscosity. The central parameter for the heat exchanger is the overall heat transfer coefficient, which could be based on either a predicted flow or most expected operational flow situations.

Various criteria for effective heat exchanger performance were established for heat transfer coefficients, perturbations in water and ammonia Reynolds numbers, average and variable flow velocities, as well as mass of flow and area of flow interface in the heat exchanger units.

Present research indicates an efficiency for OTEC closed-cycle system at 7.5 percent. However, with improvements in the design of heat exchangers and the CWP system, as well as pressure control and innovation in a direct electric grid system and indirect mode of OTEC output transportation, definite rises in the system overall efficiency and cost will emerge in the coming years.

6.7 SOLVED EXAMPLES

1. Express correlation for the predicted heat transfer coefficients and their ratio in terms of R_w and R_{NH_3} parameters. Consider average and variable fluid velocities.

Solution

$$R_w = \frac{X_w U_{w\text{-ave}}}{\nu_w} \tag{1.a}$$

$$R_{NH_3} = \frac{X_{NH_3} U_{NH_3\text{-ave}}}{\nu_{NH_3}} \tag{1.b}$$

$$U_{pred} = 0.217(R_w R_{NH_3})^{0.2345} \tag{1.c}$$

Therefore,

$$U_{pred} = 0.217 \left(\frac{X_w X_{NH_3} V_{w\text{-ave}} V_{NH_3\text{-ave}}}{\nu w \nu_{NH_3}} \right)^{0.2345} \tag{1.d}$$

The average case is

$$U_{pred} = 0.217 \left(\frac{X_w X_{NH_3} V_{w\text{-ave}} V_{NH_3\text{-ave}}}{\nu w \nu_{NH_3}} \right)^{0.2345} \tag{1.e}$$

The variable case is

$$U_{pred} = 0.217 \left(\frac{X_w X_{NH_3} W_{w\text{-var}} V_{NH_3\text{-var}}}{\nu w \nu_{NH_3}} \right)^{0.2345} \tag{1.f}$$

Therefore,

$$\frac{U_{pred\text{-ave}}}{U_{pred\text{-var}}} = \left(\frac{V_{w\text{-ave}} V_{NH_3\text{-ave}}}{V_{w\text{-var}} V_{NH_3\text{-var}}} \right)^{0.2345} \tag{1.g}$$

2. Express the overall heat transfer coefficient for the OTEC heat exchanger in terms of the fluids' internal and external inlet-outlet temperatures. Identify possible predicted and most expected cases.

Solution We have from equation 6.10

$$\Delta T = \left(\frac{2UYT}{MC_p} \right)^{1/2} \tag{2.a}$$

from which

$$U = \frac{(\Delta T)^2 MC_p}{2YT} \tag{2.b}$$

Then from equation 6.11, we have

$$\Delta T = t_1 - t_o = t_2 - t_i \tag{2.c}$$

Therefore, either

$$U = \frac{(t_1 - t_o)^2 MC_p}{2YT} \tag{2.d}$$

or

$$U = \frac{(t_2 - t_i)^2 MC_p}{2YT} \tag{2.e}$$

The predicted or most expected coefficient in either equation 2.d or 2.e depends on the time element Y and the mass of fluid flow. Therefore,

$$U_{\text{pred}} = \frac{(t_2 - t_i)^2 M_{\text{pred}} C_p}{2TY_{\text{pred}}} \tag{2.f}$$

or

$$U_{\text{most}} = \frac{(t_1 - t_i)^2 M_{\text{most}} C_p}{2tY_{\text{most}}} \tag{2.g}$$

or we can exchange by the subscript (pred) with (most) in equations 2.f and 2.g.

3. In a closed-cycle OTEC plant, express the product $R_w R_{NH_3}$ for the predicted and most expected situations with respect to internal and external fluids' inlet-outlet temperatures.

Solution We have from equation 6.2

$$U_{\text{pred}} = 0.217(R_w R_{NH_3})^{0.2345} \tag{3.a}$$

Therefore, from equation 2.f and equation 3.a,

$$R_w R_{NH_3} = \left[\frac{1}{0.217} \frac{(t_2 - t_i)^2 M_{\text{pred}} C_p}{2TY_{\text{pred}}}\right]^{4.26} \tag{3.b}$$

Similarly for the most expected case, from equations 6.3 and 2.g, we can write

$$R_w R_{NH_3} = \left[\frac{1}{0.0743} \frac{(t_1 - t_o)^2 M_{\text{most}} C_p}{2TY_{\text{most}}}\right]^{2.03} \tag{3.c}$$

4. Under what condition $R_{NH_3\text{-pred}}$ will approach infinity. Explain the operational implications of such condition in OTEC thermal exchanger unit.

Solution From equation 6.24, $R_{NH_3} \to \infty$, we can write

$$1.31(AK)^{2.9} \frac{M}{9.15YTC_p} = -1 \tag{4.a}$$

or

$$1.31(AK)^{2.9} M = 9.15YTC_p \tag{4.b}$$

where

$$K = \frac{(t_1 - t_o) - (t_2 - t_i)}{\ln (t_1 - t_o/t_2 - t_i)} \tag{4.c}$$

From equation 4.b, we can see that the contribution for $R_{NH_3} \to \infty$ may come from the following:

1. $$M = -\frac{9.15YTC_p}{1.31(AK)^{2.9}} \tag{4.d}$$

and T has to be negative temperature

2. $$AK = \left(-\frac{9.15YTC_p}{1.31M}\right)^{1/2.9} \tag{4.e}$$

3. Since

$$R_{NH_3} = \frac{X_{NH_3} Y_{NH_3}}{\nu_{NH_3}} \rightarrow \infty \tag{4.f}$$

Therefore,

$$\nu_{NH_3} \rightarrow 0 \tag{4.g}$$

5. Identify OTEC operational implications of equal normalized perturbations for R_w and R_{NH_3}.

Solution From equation 6.19, if

$$\frac{\partial R_{NH_3}}{R_{NH_3}} = \frac{\partial R_w}{R_w} \tag{5.a}$$

then,

$$\frac{20}{AK} \frac{0.217}{U_{pred}} \partial Q = 0 \tag{5.b}$$

From equation 5.b, we can imply that

1. $\partial Q = 0$, which means no change in the total rate of heat transfer.
2. $A \rightarrow \infty$, a huge interfacing area, and it is very remote situation for the heat exchanger.
3. $K \rightarrow \infty$, which implies that $t_1 - t_o = t_2 - t_i$.
4. $U_{pred} \rightarrow \infty$, which implies from equation 6.2 that either

$$R_w \rightarrow \infty$$

or

$$R_{NH_3} \rightarrow \infty$$

and hence

$$R_w R_{NH_3} \rightarrow \infty$$

6.8 REVIEW QUESTIONS

1. Compare open-cycle and closed-cycle OTEC plants with respect to system composition. Make a simple sketch of each.
2. Identify geographic world regions where it may be feasible to establish floating or land-based OTEC plants.
3. List reasons that imply that the heat exchanger is the most important component in an OTEC plant with respect to system economic effectiveness and performance.
4. Identify design difficulties for an operational CWP system for OTEC; then propose feasible solutions to those design problems.

5. Present a physical and mathematical definition for the Reynolds number for water, explaining its most effective parameter on a larger overall heat transfer coefficient.
6. Explain possible implications of a constant ocean water velocity on the average heat transfer coefficient.
7. Endothermic decomposition of water will lead to chemical mixture of H_2 and O_2. Identify the main problem of securing net amounts of H_2 and O_2. What is the nominal temperature for endothermic decomposition of water?
8. For the successful operational endothermic decomposition of water, identify an electric system for the generation of power. Write an electrochemical equation describing the release of electric current and internal electromotive force.
9. Describe a redox-flow-cell power plant that can be used to transport electric energy from an OTEC system to far destinations.
10. Establish a ratio of the U_{pred} with respect to U_{most} for a feasible OTEC system. U is the average heat transfer coefficient.

6.9 PROBLEMS

1. In an OTEC, if the predicted heat transfer coefficient is equal to the most expected value, establish a correlation between the corresponding ocean water velocities, that is, U_{pred} and U_{most}.
2. Show the systematic derivation of equation 6.19 for the difference in normalized perturbation for R_{NH_3} and R_w.
3. Equation 6.24 presents solution for R_{NH_3} in terms of OTEC system parameters. If M and, of course, A are directly linked variables, find a mathematical condition for the optimum limit of R_{NH_3}.
4. Using equation 6.24, establish the condition for optimum R_{NH_3} when Y is the only independent variable on the right-hand side of that equation.
5. Infinite value of R_{NH_3} implies zero viscosity of ammonia. Establish the mathematical criterion from equation 6.24 to that effect.
6. If the difference between the inlet internal fluid temperature and the outlet external fluid temperature is e times (the difference between the outlet internal fluid temperature and the inlet internal fluid temperature) (e = 2.718), establish solutions for Q_{pred} and Q_{most} using equations 6.14 and 6.15.
7. ΔT is expressed as $= \sqrt{2UYT/MC_p}$. Write expressions for ΔT_{ave}, ΔT_{pred}, and ΔT_{most} with respect to the evaporator unit of an OTEC heat exchanger system.
8. Express ΔT_{ave}, ΔT_{pred}, and ΔT_{most} in terms of water and ammonia variable velocities; then correlate ΔT_{ave} in terms of ΔT_{pred} and ΔT_{most}.
9. If R_w and R_{NH_3} are changing with respect to perturbations in the normalized space continuum, while the ratio of U/ν remains constant, expand U_{pred} and U_{most} using Taylor series.
10. Establish criterion for the solution of R_{NH_3} for the most expected rate of heat transfer. You may follow a procedure similar to that of obtaining equation 6.24.
11. Repeat Problem 10, but for R_w regarding the most expected rate of heat transfer.

12. An OTEC plant with ammonia as the working refrigerant is operating under a working pressure of 4 Atm for NH_3 and 1.5 Atm for both N and H_2. Solve for the prevailing R_w and R_{NH_3} in this system under the predicted case. Assume that $\gamma = 1$.

13. For Problem 12 calculate the overall ΔG in terms of γ.

14. Repeat Problem 12, but for the most expected case with respect R_w and R_{NH_3}.

15. An OTEC plant operating with ammonia at a pressure of 5 Atm and 2 Atm for nitrogen and hydrogen. Calculate the average water velocity and predicted ammonia velocity in terms of other system parameters.

16. An OTEC foam plant is operating under the criterion that energy extraction is controlled by the formula $W \approx C_p(\Delta T)^2/2T$. Express W in terms of an overall heat transfer coefficient, total time of work development, as well as M and C_p. Also, express the expected power output of this plant in terms of average water velocity, foam viscosity, and the normalized space parameters.

17. An OTEC power plant of the foam model operates in a state of adiabatic expansion; express the ratio of inlet to outlet pressures. Also solve for the total rate of thermal energy transfer if the overall heat transfer coefficient is proportional to $R_w^{1.5}$. R_w is the Reynolds number for water.

18. Using Figure 6.2, establish an empirical relationship for the overall heat transfer coefficient (U_{over}) against R_w and then against V_w. The relationship could be a piecewise linearized function. Repeat U_{over} against $\sqrt{R_w R_{NH_3}}$ from Figure 6.3; then establish a separate solution for R_{NH_3}.

19. From the results of Problem 18, establish solution for R_w.

20. With piecewise functional relationships for U versus R_w and U versus $R_w R_{NH_3}$, establish a solution for the difference in the normalized perturbations for R_w and R_{NH_3}.

21. Using Figures 6.2 and 6.3, establish piecewise functional relationships for U versus V_w and then solve for R_w and R_{NH_3} together with their normalized perturbations.

22. Present several physical and engineering operating conditions under which $R_w = R_{NH_3} \rightarrow \infty$. Repeat for $R_w = R_{NH_3} \rightarrow 0$. Relate the effect of the mentioned values of R_w and R_{NH_3} on their flow velocities.

6.10 REFERENCES

1. Bogardus, H., D. A. Kruger, and D. Thompson. "Dynamic Magnetization in Ferrofluids." *Journal of Applied Physics,* Vol. 49, no. 6 (June 1978), pp. 3422–3429.
2. Denno, K. "Hybrid System of OTEC Plantship Couple to MHD and Hydrogen Fuel-Cell Generating Plants." *Alternative Energy Sources IV,* Vol. 5, edited by T. Nejat Veziroglu, pp. 239–251. Ann Arbor Science Publishers, Michigan, 1982.
3. Denno, K. "Ocean Thermal Energy Conversion, Storage and Transport Through Water-Based Ferric Fluid," *Alternative Energy Sources IV,* Vol. 4,

edited by T. Nejat Veziroglu, pp. 55–64. Ann Arbor Science Publishers, Michigan, 1982.

4. Denno, K. "Modelling of Energy Release Systems from OTEC Plants." *Alternative Energy Sources IV,* Vol. 3, edited by T. Nejat Veziroglu, pp. 141–149. New York: Hemisphere Publishing Corporation, 1985.
5. Denno, K. "Magnetic Transport Properties of Dipolar Conducting Liquids at Ambient Temperatures." *Journal of Electrostatics,* Vol. 7 (1979), pp. 345–360.
6. Denno, K. "Mathematical Modelling of Storage Batteries and Fuel Cell." *International Telephone Energy Conference,* 78CH1353-2, INTELEC 1978, pp. 237–243.
7. Ford, G., C. Niblett, and L. Walker. "Ocean Thermal Energy Conversion." *IEE Proceedings,* Vol. 130, Pt. A, no. 2 (March 1983), pp. 93–100.
8. Pandolfini, P., G. L. Dugger, F. K. Hill, W. H. Avery, and H. D. Faust. "AL Clad Aluminum, Folded-Rube Heat Exchangers for OTEC." *Proceedings of Condensed Papers,* 3rd Miami International Conference on Alternative Energy Sources, Dec. 1980 pp. 424–426.
9. Stoecker, W. F. *Design of Thermal Systems.* New York: McGraw-Hill Book Co., 1971.
10. Velingker, A., K. Sohn, and K. Denno. "More on the Parametric Transformation of Water-Based Ferro-Magnetic Solutions for Energy Conversion." *Alternative Energy Sources VI,* Vol. 3, edited by T. Nejat Veziroglu, pp. 321–334. New York: Hemisphere Publishing Corporation, 1985.
11. Warshay, Marvin, and L. O. Wright. "Cost and Size Estimates for an Electrochemical Bulk Energy Storage Concept." *NASA Technical Memorandum,* NASA TM X-3192. Lewis Research Center, Cleveland, Ohio, February 1975.
12. Thaller, Lawrence. "Electrically Rechargeable Redox Flow Cells." *NASA Technical Memorandum,* TM-X-71540. Lewis Research Center, Cleveland, Ohio, August 1974.

7

Linkage of OTEC to Other Modes of Energy Systems

This chapter presents concepts of topping systems of energy release with respect to linkage to ocean thermal energy conversion (OTEC) plantship. Modes of energy release include the utilization of the bulk ΔT from the huge volume of ocean water for endothermic decomposition and conventional electrolysis of water into hydrogen and oxygen. The decomposition and electrolysis processes have to be followed by separation cycle to secure pure hydrogen and oxygen.

Air liquefaction coupled with the availability of hydrogen could produce ammonia NH_3, which may be recycled in OTEC or used in agricultural purposes. The release of hydrogen and oxygen could be used as fuel for the conventional fuel cell leading to the generation of electric power. Another applicable mode of energy conversion is the linkage of magnetohydrodynamic (MHD) generating plants to OTEC. The huge bulk ΔT from OTEC will be used effectively to raise the temperature of the working fluid to the desired optimum level for efficient MHD power generation.

Next utilization of OTEC–ΔT output could be used to produce steam-powering turbines, working as prime mover to generate conventional electric power. Part or all of the power produced by conventional plant could be utilized to recharge a redox-flow-cell energy system that uses $FeCl_2$ for the catholyte and $TiCl_3$ in the anolyte. The recharging cycle will transform the ferrous solution in the catholyte into ferric solution, which after a substantial increase in its concentration, may be transported to any destination where it could be reused again after certain dilution as the working fluid for the catholyte of the redox flow cell. Therefore, the linkage of the redox flow cell to OTEC through the conventional power plant will serve the purpose of storing OTEC output in a chemical continuum and transporting it to the desired location.

Another concept of using OTEC through system interconnection is through the utilization of oxygen released by the endothermic decomposition and conventional electrolysis of water as oxidizing agent for converting prepared refuse into an ordered hydrocarbon, namely, C_mH_{2m+2} and then a follow-up cycle to oxidize the ordered synthetic fuel to produce electric power through the function of the bioelectrochemical fuel cell.

7.1 HYBRID SYSTEM OF OTEC PLANTSHIP COUPLED TO MHD AND HYDROGEN FUEL-CELL GENERATING PLANTS[1]

7.1.1 Introduction

Present modes of thermal energy extraction from oceans in tropical zones call for the production of ammonia through the mixture of nitrogen obtained from liquefied air and hydrogen secured by separation from the endothermic decomposition and/or conventional electrolysis of water. Consequently, ammonia will be transported for demand zones either in gaseous or liquid state.

This section presents, first, a new mode of on-site utilization of ocean thermal energy as an effective source for the production of working conducting fluid in MHD electric power generation by preheating the natural gas to the required thermal operational level. The second mode is the generation of electric power by conventional fuel cell, in which hydrogen secured through water decomposition will serve as the primary fuel input while air heated by ocean thermal energy will act as the oxidizer.

The two modes of electric power generation, namely, the MHD and the fuel-cell systems, could be set up either individually or jointly as a hybrid system directly coupled to an off-shore OTEC plantship or on a land-based site very close to the OTEC plant location.

Calculations have been carried out for the similation of a hybrid system interconnecting an off-shore OTEC with MHD plants and through electric power inverter to deliver an alternating power that could be delivered to far load centers through high-tension transmission lines. The OTEC-MHD-inverter system simulation presents compatible relationships among the

[1] © Material and figures in Section 7.1 is reprinted, with permission, from Alternative Energy Sources IV, Vol. 5, pp. 239–251, 1982. "Hybrid System of OTEC Plantship Coupled to MHD and Hydrogen Fuel Cell Generating Plant" by K. Denno. Edited by T. Nejat Veziroglu. 1982. Ann Arbor Science Publishers.

OTEC parametric performance elements, namely, the total as well as the average heat rate collection, water velocity, and Reynolds number for water, on one hand, and those corresponding values of the MHD plant, which include plasma temperature, pressure, and velocity as well as plant loading factor, actual output power, efficiency, and the AC inverter commutation characteristics, on the other.

The next stage of work in this section is the dynamic model simulation of an interconnected hybrid system of MHD–hydrogen–fuel-cell, a DC electric power generating system coupled to an offshore OTEC plant and the delivery of AC output through auxiliary impulse commutated power inverters. Compatibility analysis for the OTEC–MHD–fuel-cell system has been carried out with the development of operational criteria for optimum outputs, efficiency, and loading conditions of the MHD–fuel-cell system with respect to OTEC plant parametric energy characteristics. Also a coordination planning relationship is established for OTEC thermal energy distribution with respect to hydrogen production, MHD plasma heating, and air preheating for oxidation function in fuel-cell operation.

The simulation presented in this text adds to the effective development and applications of the OTEC system and other modes of alternative energy sources discussed earlier.

7.1.2 System Simulation of Water Endothermic Decomposition and Conventional Electrolysis with Air Liquefaction

Figure 7.1 shows a schematic interconnection of an OTEC plant for generation of H_2, O_2, and NH_3. With respect to this figure,

Q_1, Q_2 = waste heat energy output of OTEC and endothermic energy conversion processes

E_0 = OTEC bulk energy input

E_1 = net thermal energy extracted from OTEC toward an input of endothermic process

E_2 = energy content out of H_2 and O_2 prior to the separation process

E_3 = OTEC energy output for air liquefaction process

$Q_3 = Q_1 + Q_2$

Q_4 = energy content for H_2 and O_2 out of conventional electrolysis for H_2 and O_2

ΔO_D, ΔO_C = molar enthalpy change for oxygen in the endothermic and electrolysis process, respectively

ΔH_D, ΔH_C = hydrogen enthalpy change in endothermic and conventional electrolysis of water, respectively

ΔN_L = molar nitrogen enthalpy change through the process of air liquefaction

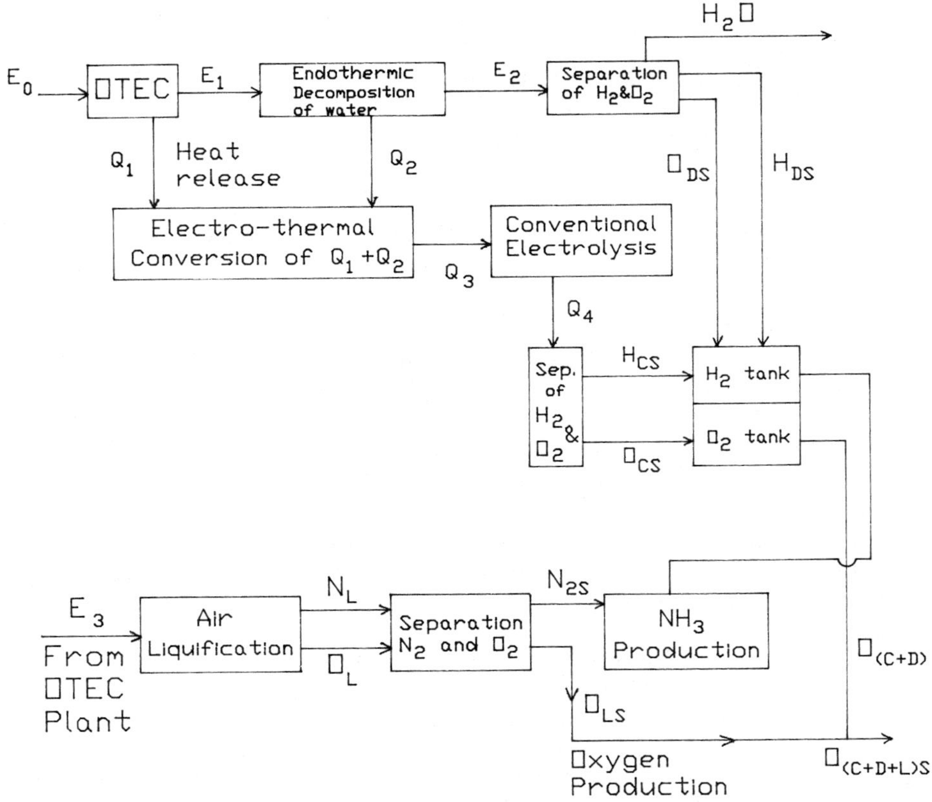

Figure 7.1 Endothermic and conventional electrolysis of water for ammonia production, hydrogen, and oxygen.

ΔN_S = molar nitrogen enthalpy change in the process of separation

ΔO_L = molar oxygen enthalpy change for O_2 out of air liquefaction process

ΔO = molar oxygen enthalpy change for O_2 out of H_2, O_2 separation process

$\Delta O_{(C+D)}$ = total molar oxygen enthalpy change out of electrolysis and endothermic decomposition of water

$\Delta H_{(C+D)}$ = total molar hydrogen enthalpy change out of electrolysis and endothermic decomposition of water

$\Delta H_{(C+D)S}$ = total molar hydrogen enthalpy change out of electrolysis-endothermic and separation process

ΔO_{LS} = total oxygen enthalpy change out of air liquefaction and separation process

$\Delta O_{(C+D+L)S}$ = total oxygen enthalpy change out of electrolysis, endothermic, air liquefaction, and separation processes

Q_3 = electrical output of $Q_1 + Q_2$ heat input released
n_{HD} = the molar concentration for hydrogen out of endothermic decomposition process
n_{OCS} = the molar concentration of oxygen out of electrolysis-separation processes
n_{HCS} = the molar concentration of hydrogen out of electrolysis-separation processes
n_{HDS} = the molar concentration of hydrogen out of endothermic-separation processes
n_{NL} = the molar concentration of nitrogen out of air liquefaction process
n_{NLS} = the molar concentration of nitrogen out of the process of separation
n_{OL} = the molar concentration of oxygen out of air liquefaction
n_{OLS} = the molar concentration of nitrogen out of the process of separation

7.1.3 Release of Hydrogen Through Endothermic Decomposition and Conventional Electrolysis of Water

Let us list the following plant efficiencies based on what is shown in Figure 7.1

$$\eta_1 = \frac{E_1}{E_0} \tag{7.1}$$

$$\eta_2 = \frac{E_2}{E_1} \tag{7.2}$$

$$\eta_3 = \frac{\mathrm{O}_{DS}\mathrm{n}_{ODS} + H_{DS}\mathrm{n}_{HDS}}{E_2} = \frac{\Delta\mathrm{O}_{DS}\mathrm{n}_{OS}}{E_2} + \frac{\mathrm{H}_{DS}\mathrm{n}_{HDS}}{E_2} \tag{7.3}$$

$$\eta_4 = \frac{Q_3}{Q_1 + Q_2} \tag{7.4}$$

$$\eta_5 = \frac{Q_4}{Q_3} \tag{7.5}$$

$$\eta_6 = \frac{\Delta\mathrm{H}_{CS}\mathrm{n}_{HCS} + \Delta O_{CS}\mathrm{n}_{OCS}}{Q_4} \tag{7.6}$$

We may express the following molar concentrations based on the structure of the reaction equation:

n_{OC} = the molar concentration for oxygen out of conventional electrolysis
n_{OD} = the molar concentration for oxygen out of the endothermic decomposition process
n_{HC} = the molar concentration of hydrogen out of conventional electrolysis

From equation 7.3, let

$$n_{ODS} = \frac{\mathrm{O}_{ODS} n_{ODS}}{E_2} \tag{7.7}$$

which is the efficiency of oxygen separation out of the endothermic process. Therefore, the efficiency of hydrogen separation following the endothermic process identified by η_3 is expressed by

$$\eta_{HDS} = \frac{\Delta \mathrm{H}_{DS} n_{HDS}}{E_2} \tag{7.8}$$

$$= \eta_3 - \mathrm{O}_{DS} \tag{7.9}$$

or

$$n_{HDS} = \frac{E_2(\eta_3 - \eta_{ODS})}{\Delta \mathrm{H}_{DS}} \tag{7.10}$$

$$\mathrm{H}_{DS}\mathrm{H}_{DS} = \eta_1 \eta_2 E_0(\eta_3 - \eta_{ODS}) \tag{7.11}$$

And by knowledge of hydrogen enthalpy change H_{DS}, which may include experimental data, the molar concentration of hydrogen released by endothermic process is expressed by*

$$n_{HDS} = \frac{\eta_1 \eta_2 E_0(\eta_3 - \eta \mathrm{O}_{DS})}{\mathrm{H}_{\mathrm{DS}}} \tag{7.12}$$

Similarly, hydrogen release due to conventional electrolysis can be expressed in the form of its molar concentration. Let

$$\eta_{OCS} = \frac{\Delta \mathrm{O}_{CS} + n_{OCS}}{Q_4} \tag{7.13}$$

and

$$\eta_{HCS} = \frac{\Delta \mathrm{H}_{CS} n_{HCS}}{Q_4} \tag{7.14}$$

Therefore,

$$\eta_{HCS} = \eta_6 - \eta_{OCS} \tag{7.15}$$

$$= \frac{\Delta \mathrm{H}_{CS} n_{HCS}}{Q_4} \tag{7.16}$$

or

$$\eta_{HCS} = \frac{\eta_4\eta_5(Q_1 + Q_2)[\eta_6 - \eta_{OCS}]}{\Delta H_{CS}} \tag{7.17}$$

By now, we can write implicitly

$$n_{H_{(C+D)S}} = n_{HDS} + n_{HCS} = \frac{\eta_1\eta_2E_0(\eta_3 - \eta_{ODS})}{H_{DS}} + \frac{\eta_4\eta_5(Q_1 + Q_2)(\eta_6 - \eta_{OCS})}{\Delta H_{CS}} \tag{7.18}$$

In equation 7.18 the molar enthalpy change for hydrogen in the process of endothermic with separation and conventional electrolysis with separation, respectively, must be secured either experimentally or by any feasible modeling procedure.

However, if $\Delta H_{CS} = \Delta H_{DS} = H_{CDS}$, which may be visualized through modeling, we can express the total molar concentration of hydrogen produced by endothermic decomposition and conventional electrolysis of water as followed by separation as

$$H_{(C+D)S} = n_{HDS} + n_{HCS} = \frac{\eta_1\eta_2E_0(\eta_3 - \eta_{ODS}) + \eta_4\eta_5(Q_1 + Q_2)(\eta_6 - \eta_{OCS})}{\Delta H_{CDS}} \tag{7.19}$$

7.1.4 Extraction of Oxygen Through Endothermic Decomposition and Conventional Electrolysis of Water

Efficiency of oxygen extraction from endothermic decomposition of water followed by separation is expressed by

$$\eta_{ODS} = \frac{\Delta O_{DS} n_{ODS}}{E_2} \tag{7.20}$$

$$= \eta_3 - \eta_{HDS} \tag{7.21}$$

or

$$n_{ODS} = \frac{E_2(\eta_3 - \eta_{HDS})}{\Delta O_{DS}} \tag{7.22}$$

and, of course,

$$n_{ODS}O_{DS} = \eta_1\eta_2E_0(\eta_3 - \eta_{HDS}) \tag{7.23}$$

Hence by knowledge of external change in oxygen molar enthalpy O_{DS} out from the process of endothermic decomposition and the followed separation, we can write,

$$n_{ODS} = \frac{\eta_1\eta_2E_2(\eta_3 - \eta_{HDS})}{\Delta O_{DS}} \tag{7.24}$$

Oxygen extraction from conventional electrolysis may follow a path similar to that for hydrogen, where,

$$\eta_{OCS} = \eta_6 - \eta_{HCS} \tag{7.25}$$

$$= \frac{\Delta O_{CS} n_{OCS}}{Q_4} \tag{7.26}$$

or

$$n_{OCS} = \frac{\eta_4 \eta_5 (Q_1 + Q_2)(\eta_6 - \eta_{HCS})}{\Delta O_{CS}} \tag{7.27}$$

where O_{CS} is the oxygen excess change in molar enthalpy due to the endothermic and conventional electrolysis processes.

Implicitly the total molar concentration of extracted oxygen from endothermic decomposition of water and conventional electrolysis can be expressed by

$$n_{O(C+D)S} = n_{ODS} + n_{OCS} = \frac{\eta_1 \eta_2 E_0 (\eta_3 - \eta_{HDS})}{O_{DS}} + \frac{\eta_4 \eta_5 (Q_1 + Q_2)(\eta_6 - \eta_{HDS})}{\Delta O_{CS}} \tag{7.28}$$

7.1.5 Ammonia Formation

Release of ammonia can be realized with the chemical union of hydrogen produced by the processes of endothermic decomposition and/or conventional electrolysis of water and nitrogen that can be secured through air liquefaction. Again, referring to Figure 7.1, let

$$\eta_{ONL} = \frac{n_{OL} \Delta O_{2L} + n_{NL} \Delta N_L}{E_3} \tag{7.29}$$

where η_{ONL} is the efficiency of oxygen, nitrogen release out of air-liquefaction process. Also, we can testify η_{ONLS} as the efficiency of oxygen and nitrogen release followed by the process of separation. η_{ONLS} can be written as

$$\eta_{ONLS} = \frac{n_{OLS} \Delta O_{2LS} + n_{NLS} \Delta N_{LS}}{n_{OL} \Delta O_{2L} + n_{NL} \Delta N_{NL}} \tag{7.30}$$

$$= \frac{n_{OLS} \Delta O_{2LS} + n_{NLS} \Delta N_{LS}}{\eta_{ONL} E_3} \tag{7.31}$$

$$= \frac{n_{OLS} \Delta O_{2LS}}{\eta_{ONL} E_3} + \frac{n_{NLS} \Delta N_{LS}}{\eta_{ONL} E_3} \tag{7.32}$$

Let η_{OLS} be the efficiency of oxygen extraction air liquefaction followed by the process of separation;

$$\eta_{OLS} = \frac{n_{OLS}\Delta O_{2LS}}{\eta_{ONL}E_3} \tag{7.33}$$

Therefore,

$$\eta_{NLS} = \eta_{ONS} - \frac{n_{NLS}\Delta N_{LS}}{\eta_{ONL}E_3} \tag{7.34}$$

Hence, we can write for nitrogen release out of air-liquefaction and separation processes

$$n_{NLS}\Delta N_{LS} = (\eta_{OLS} - \eta_{NLS})\eta_{ONL}E_3 \tag{7.35}$$

Now we can see from equations 7.9 and 7.35 the possibility of structuring the molar concentration of ammonia as expressed below, by compounding proportioned ratios,

$$k_1 n_{H(C+D)S} + k_2 n_{NLS} = k_3 \left[\frac{\eta_{ONL}E_3(\eta_{OLS} - \eta_{NLS})}{\Delta N_{LS}} + \frac{\eta_1\eta_2 E_o(\eta_3 - \eta_{ODS}) + \eta_4\eta_5(Q_1 + Q_2)(\eta_6 - \eta_{OCS})}{\Delta H_{CDS}} \right] \tag{7.36}$$

However, for compounding of ammonia NH_3, the ratio of concentration by weight of

$$\frac{n_{NLS}}{n_{H(C+D)S}} = \frac{14}{3}$$

Therefore,

$$\begin{aligned} k_1 &= 3 \\ k_2 &= 14 \\ k_3 &= 17 \end{aligned} \tag{7.37}$$

It is useful at this stage to reiterate that N_{LS}, H_{CDS} represents the excess or change in molar enthalpy for nitrogen extracted by air-liquefication–separation and hydrogen secured through conventional electrolysis and endothermic decomposition–separation processes, respectively.

Another way to determine the change in molar enthalpy for either H_2 or N_2 may be through linkage according to the following relationship,

$$aP_N + bP_H \rightarrow CP_{NH_3} \tag{7.38}$$

where

a, b, c = the molar concentration in the reaction equation
P_N, P_H, P_{NH_3} = the operating pressure or activity for nitrogen, hydrogen, and ammonia, respectively

The terms a, b, and c can be extracted from the following balanced chemical equation:

$$2NH_3 \rightarrow N_2 + 3H_2 \tag{7.39}$$

Therefore,

$$c = 2$$

$$b = 3$$

$$a = 2$$

In terms of the Gibbs free energy function, equation 7.38 may be written as

$$\Delta G - G_0 = RT \ln \frac{P_N^a P_{H_2}^b}{P_{NH_3}^c} \tag{7.40}$$

where

ΔG_0 = the perturbation of Gibbs function at standard pressure and temperature
ΔG = the perturbation in Gibbs function at any pressure and temperature

We may provide linkage of equation 7.40 to water and ammonia Reynolds numbers as indicated in equation 6.29, where

$$\Delta G = G - G_0 = \gamma(R_w R_{NH_3})^{0.2345} \tag{7.41}$$

As indicated there, R_w and R_{NH_3} are functions of the average and variable water velocities as well as the average and the overall heat transfer coefficients.

7.1.6 Linkage of OTEC (H_2-O_2) Release to Conventional Fuel Cell

As presented earlier in Chapter 1, the electromotive force (emf) produced by H_2-O_2 cell is expressed by the Nernst equation as

$$e = e_0 + \frac{RT}{n\Gamma} \ln \frac{P_{H_2O}^c}{P_{H_2}^a P_{O_2}^b} \tag{7.42}$$

where

$$O_2 + 2H_2 \rightarrow 2H_2O \tag{7.43}$$

$P_{H_2}, P_{O_2}, P_{H_2O}$ = the pressure or activity of H_2, O_2, and H_2O, respectively
a, b, c = the molar concentration for H_2, O_2, and H_2O, respectively, as indicated in equation 7.42
e_0, e = the cell emf at standard pressure and temperature and at any pressure and temperature, respectively.

If we consider that a system of interconnected fuel cells supplies a total power of VI watts for a time duration of t seconds and has an operational efficiency of η_F, then we can express the total energy input to the fuel-cell system from OTEC plant supplying H_2 and O_2:

$$W = \frac{VIt}{\eta_F} \text{ J} \tag{7.44}$$

or taking into consideration the total concentration for oxygen and hydrogen extracted from OTEC system,

$$n_{O(C+D)S} + n_{H[C+D]S} = n_{O\text{-}H_{[C+D]S}} \tag{7.45}$$

Therefore, where the subscript $[C+D]S$ implies conventional electrolysis, endothermic decomposition is followed by separation.

The differential enthalpy or enthalpy external change for hydrogen and oxygen combined (O-H) can be expressed as

$$O_{[C+D]S} + H_{[C+D]S} = (\text{O-H})_{[C+D]S} \tag{7.46}$$

$$(\text{O-H})_{[C+D]S} = \frac{VIt}{\eta_F n_{\text{O-H}_{[C+D]S}}} \tag{7.47}$$

Equation 7.46 provides a useful vehicle to secure information for either the total molecular differential enthalpy or total molecular concentration for the separated oxygen and hydrogen obtained by the endothermic and conventional electrolysis that is energized by thermal energy produced from OTEC system.

Linkage of a fuel-cell system to an OTEC plant, for example, may be realized with that of the foam system, from equation 6.30 energy expansion produced is rewritten as

$$W \approx \frac{C_p(\Delta T)^2}{2T}$$

7.1.7 OTEC-MHD Connection

T generated by an OTEC plantship in a tropical ocean zone through an isentropic foam system linked to condenser and/or evaporator components can be used effectively in raising MHD plasma temperature to a level very compatible for MHD generator working fluid that is about 4000°F. Enhancement of working conditions in the MHD power plant is demonstrated through changes in the following MHD relationships, the complete set of MHD equations containing the subset of flow field and Maxwell field equations.

Flow Field Equations

The Continuity Equation

$$\frac{D\rho}{Dt} + \rho\overline{\nabla} \cdot \overline{V} = 0 \tag{7.48}$$

where the convective derivation D/Dt is expressed by

$$\frac{D}{Dt} = \frac{\partial}{\partial t} + \overline{V} \cdot \overline{\nabla} \tag{7.49}$$

For an incompressible working fluid, $\overline{\nabla} \cdot \overline{V} = 0$.

Momentum Equation. For an arbitrary volume of MHD working fluid, incompressible and of zero viscosity, the momentum equation is expressed by

$$(\overline{V} \cdot \overline{\nabla})\overline{V} = -\nabla P + \overline{J} \times \overline{B} \tag{7.50}$$

Energy Equation. Again for MHD working fluid is incompressible and of zero viscosity, the energy equation is

$$\rho(\overline{V} \cdot \overline{\nabla}) \left[\frac{1}{2} v^2 + \frac{P}{\rho} + C_v \Delta T \right] = \overline{E} \cdot \overline{J} \tag{7.51}$$

where in the flow field equations

V = the velocity vector of the working fluid
ρ = the mass density in kilograms per cubic meter
T = the absolute temperature of the working fluid in kelvins
C_v = the working fluid specific heat at constant volume
E = the total electric field that includes any applied and the induced fields

Therefore,

$$\overline{E} = \overline{E}_{ind} + \overline{E}_a \tag{7.52}$$

where

E_{ind} = the induced electric field in volts per meter
E_a = any applied electric field
J = the induced current density vector in amperes per square meter

Maxwell Electromagnetic Field Equations

The Charge Continuity Equation

$$\overline{\nabla} \cdot \overline{J} + \frac{\partial \rho_e}{\partial t} = 0 \tag{7.53}$$

where ρ_e is the net charge density.

Ampere's Law

$$\overline{\nabla} \times \overline{B} = {}_0\mu \overline{J}_t \tag{7.54}$$

$$\overline{\nabla} \cdot \overline{B} = 0 \tag{7.55}$$

Faraday's Equation

$$\overline{\nabla} \times \overline{E} = -\frac{\partial \overline{B}}{\partial t} \tag{7.56}$$

where

B = the total magnetic induction in tesla

${}_0\mu$ = the magnetic permeability for nonmagnetic medium measured in henries per meter and is equal to $4\pi \times 10^{-7}$ henry/m

Ohm's Law

$$\overline{J}_c = \sigma(\overline{E}_a + \overline{V} \times \overline{B}) \tag{7.57}$$

where σ is the electrical conductivity in ohm-meters.

The Energy Equation

$$W_{em} = \overline{E}_a \cdot \overline{J}_t - \frac{J_c^2}{\sigma} \tag{7.58}$$

where

J_t = the total current density, that is, $\overline{J}_t = \overline{J}_c + \overline{J}_D$

J_c = arc conduction current density

J_D = the convective current density, which is represented by

$$\overline{J}_D = \partial \overline{D} / \partial t \tag{7.59}$$

D = the electric displacement measured in coulomb per square meter = εE, where ε is the working fluid permittivity in farads per meter

Direct input of OTEC to an MHD electric power generating system is seen by looking at the entire MHD set of equations that could be accomplished through the element of temperature (T) of the working fluid as indicated in equation 7.50.

Raising the working fluid temperature will enhance its electrical conductivity σ and hence the induced conductive current density as indicated by equation 7.56 and eventually the MHD energy developed as shown in equation 7.57.

Now, we may present a special case regarding combustion plasma gas, with 1 percent seeding potassium, that through piecewise linearization, the electrical conductivity σ versus T could be listed as shown:

For:

1. $1400 < T < 1700°F$

$$\sigma_1 = \tfrac{2}{15} T - 187 \quad (7.60)$$

2. $1700 < T < 2500°F$

$$\sigma_2 = \tfrac{9}{20} T - 725 \quad (7.61)$$

3. $2500 < T < 5000°F$

$$\sigma_3 = \tfrac{1}{25} T + 2000 \quad (7.62)$$

Hence, we can see from equations 7.60, 7.61, and 7.62 the following incremental temperatures in each of the respective three operating regions,

$$\Delta T_1 = \tfrac{2}{15} T_1, \text{ in region 1} \quad (7.63)$$

$$\Delta T_2 = \tfrac{9}{20} T_2, \text{ in region 2} \quad (7.64)$$

and

$$\Delta T_3 = \tfrac{1}{25} T_3, \text{ in region 3} \quad (7.65)$$

Therefore,

$$\Delta\sigma \propto \Delta T \quad (7.66)$$

A systematic block diagram that illustrates the linkage of OTEC thermal output in the form of bulk ΔT to the processes of endothermic decomposition of water into H_2 and O_2 as well as through conventional electrolysis, providing inputs to a fuel-cell system and also the linkage of the remaining part of ΔT to enhance the working fluid temperature of MHD power generation system, is shown in Figure 7.2.

7.2 OTEC STORAGE AND TRANSPORT THROUGH WATER-BASED FERRIC FLUID

7.2.1 Introduction[2]

This section presents a feasibility study of an expanded system for the storage and transport of ocean thermal energy through a simulation system encompassing storage in another chemical form, namely, the water-based ferric fluid output from the catholyte of the redox flow cell throughout the process of recharging. The new transform to ferric liquid may be the unique mode of ocean thermal energy conversion or may constitute a supplement to the production of ammonia. Generation of ferric liquids is accomplished by

[2] Denno, ''Hybrid System of OTEC. . . ''

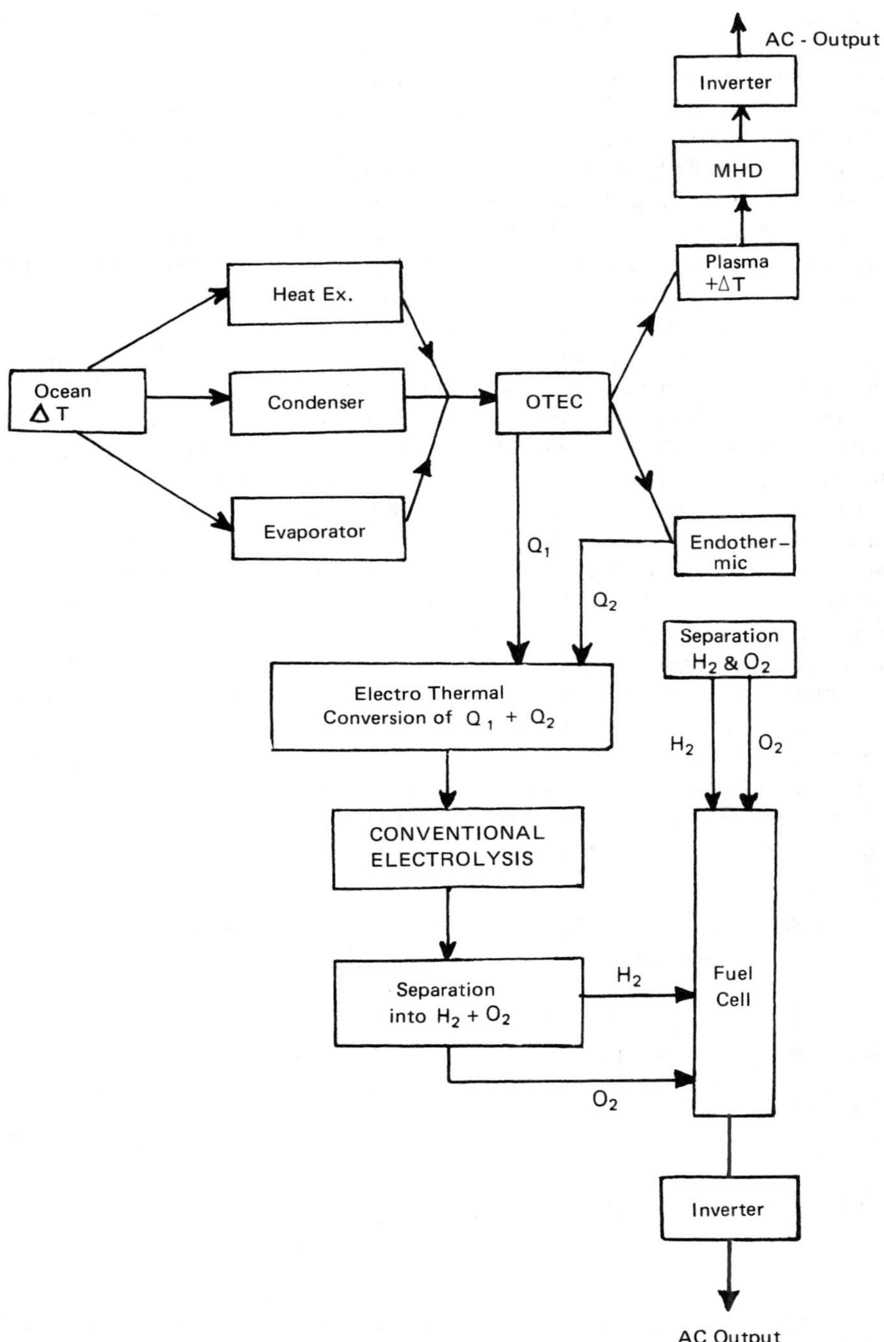

Figure 7.2 OTEC coupling to MHD and fuel-cell generators.

the initial step of transforming the oceanic thermal energy to an electrical form, which, in turn, is injected through the redox-flow-cell system in a charging process designed to store the electrical DC power in a chemical form by converting the catholyte ferrous solution into a ferric water-based fluid. The ferric liquid could be transported in oxidation-free tanks to the consuming centers that will receive the electrical energy supplied from generating redox-flow-cell systems, with the ferric solution storing the oceanic thermal energy and serving as the catholyte working fluid.

Simulation model has been developed earlier for the generation of oceanic thermal energy represented by the average heat collection and its dependence on the seawater Reynolds number as well as water velocity, conversion of thermal energy to electrical output, and then the process of recharging the redox-flow-cell plant to store the electrical energy in chemical form through transforming a ferrous field in the catholyte to a ferric one. A control system guiding the simulation model is established to ensure stability and continuity for compatible coordination between the process of oceanic heat production and electrical output with respect to the recharging stage of the redox system. The stability aspect of the entire model has been analyzed with respect to performance parametric properties of the redox flow system, with the assumption that the effects of its internal circulating current are completely neutralized; hence no adverse limitation will be imposed on its input electrical potential range. Following the simulation of energy storage in ferrous liquid form, an overall system of thermal energy production, storage, and transport has been established, including ammonia as well as ferrous solution with hydrogen generation secured concurrently through the endothermic decomposition and conventional electrolysis of water.

7.2.2 Recharge Performance Equation of the Redox Flow Cell[3]

For the ferromagnetic catholyte solution and titanium anolyte of the redox flow cell, we may rewrite the basic electrochemical and balanced energy equations relevant to the presentation introduced in Chapter 4.

For the catholyte,

$$\mathrm{Fe}^{+2} \rightarrow \bar{e} + \mathrm{Fe}^{+3}$$

And, in terms of reactants activities or partial pressures,

$$P^{a}_{\mathrm{Fe}^{+2}} \rightarrow P^{b}_{\bar{e}} + P^{c}_{\mathrm{Fe}^{+3}}$$

[3] Denno, "Ocean Thermal Conversion, Storage and Transport through Water-Based Ferric Fluid," *Alternative Energy Sources IV,* Vol. 4, pp. 55–64, 1982; Ann Arbor Science Publishers.

where a, b, and c express the molar concentration of Fe^{+2}, $\bar{e}$ (electron cloud), and Fe^{+3}, respectively, while P represents the corresponding activity or partial pressure.

Then in terms of the Gibbs free energy function and its reference generating emf, the oxidizing differential voltage is derived as follows:

$$\Delta E_{Fe} = \frac{RT_{Fe}}{F\,n} \ln \frac{P_{Fe^{+2}}}{P_{\bar{e}} P_{Fe^{+3}}} \tag{7.67}$$

where

R = universal gas constant
T = operating reaction temperature
F = Faraday's constant
n = number of electron moles released

Similarly for the anolyte chamber reaction,

$$Ti^{+4} + \bar{e} \rightarrow Ti^{+3}$$

and, hence,

$$\Delta E_{Ti} = \frac{RT_{Ti}}{nF} \ln \frac{P_{Ti^{+4}} P_{\bar{e}}}{P_{Ti^{+3}}} \tag{7.68}$$

and the total reaction equation becomes

$$FeCl_3 + TiCl_3 \rightarrow FeCl_2 + TiCl_4$$

Therefore,

$$E_{total} = \frac{RT}{nF} \ln \frac{P_{FeCl_3} P_{TiCl_3}}{P_{FeCl_3} P_{TiCl_4}} \tag{7.69}$$

Equation 7.69 is valid if $T_{Fe} = T_{Ti} = T$ and

$$\Delta E_{Fe} = \Delta E_{Ti} = E_{total} \tag{7.70}$$

General criterion for recharge is

$$P_{Ti^{+3}} P_{Fe^{+2}} = P_{\bar{e}}^2 P_{Ti^{+4}} P_{Fe^{+3}} \tag{7.71}$$

The recharge electric input power becomes

$$E_{rech} = JA \frac{RT}{F} \ln \frac{P_{FeCl_3} P_{TiCl_3}}{P_{FeCl_2} P_{TiCl_4}} \tag{7.72}$$

where

J = redox cell current density in amperes per square meter
A = electrode area

At the place of destination, ferric solution containing the stored electrochemical energy transformed throughout recharging will be used in the catholyte chamber for the reverse process of discharge and the delivery of electric power to the load.

7.2.3 Discharge Equations in the Redox Cell[4]

Relevant to the presentation in Chapter 4 on the redox flow cell, it is useful to rewrite the basic discharge and balanced energy equations of the redox flow cell as indicated below:

At the catholyte,

$$FeCl_3 + \bar{e} \rightarrow FeCl_2 + Cl^-$$

$$P_{FeCl_3} P_{\bar{e}} = P_{FeCl_2} P_{Cl^-} \tag{7.73}$$

And

$$E_g - E_{g_0} = \Delta E_{g_{out}} = \frac{RT}{n\mathrm{F}} \ln \frac{P_{FeCl_3} P_{\bar{e}}}{P_{FeCl_3} P_{Cl^-}} \tag{7.74}$$

where

E_g = the generated emf under any level of activity or partial pressure
E_{g_0} = the generated emf under standard level of pressure and temperature

At the anolyte,

$$P_{TiCl_3} P_{Cl^-} = P_{TiCl_4} P_{\bar{e}} \tag{7.75}$$

and

$$\Delta Eg_{out} = \frac{RT}{n\mathrm{F}} \ln \frac{P_{TiCl_3} P_{Cl^-}}{P_{TiCl_4} P_{\bar{e}}} \tag{7.76}$$

A simulation diagram for the systematic OTEC-redox system is shown in Figure 7.3. The system operates in a rich temperature gradient oceanic zone with an off-shore or floating OTEC plant. The sequential power conversion and extraction operations function on the same OTEC site or in close proximity. The terminal stage of operation includes the redox-flow-cell power system whose electric DC output has to be linked to a power solid-state inverter for AC power output and then followed by the required stepping-up transformer.

Some useful and timely comments on the aspects of an OTEC system's contribution to energy storage, transport, and utilization are as follows:

[4] Ibid.

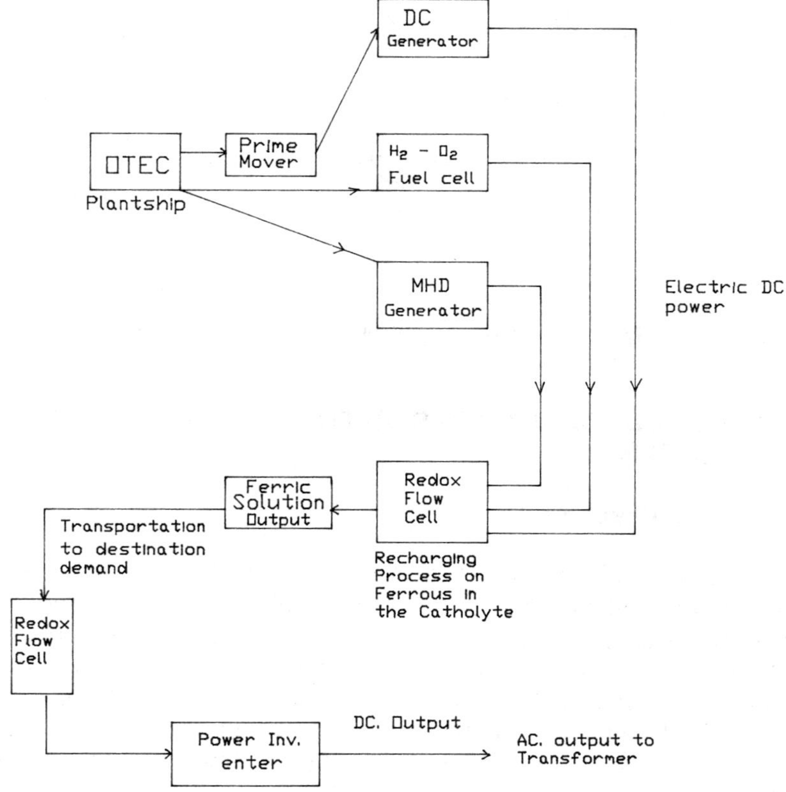

Figure 7.3 OTEC-redox storage system. (© 1982 Ann Arbor Science Publishers.)

1. Electrochemical feasibility demonstrates the aspect for conversion of OTEC energy output in ferric water-based liquids which can be transported to demand centers for electrical power.
2. Conversion of ferrous water-based solutions of several molar concentrations to ferric liquids is secured in a recharging process of the redox flow cells by applying electric DC voltage across the anolyte-catholyte terminals.
3. Criteria have been presented regarding conventional modes of ocean thermal energy conversion by heat exchangers, condensers, and evaporators in terms of water flow velocity, water Reynolds number, as well as the coefficient of heat transfer.
4. Conventional as well as advanced means of transforming OTEC output to electrical form can be employed on the same site of OTEC plantship.

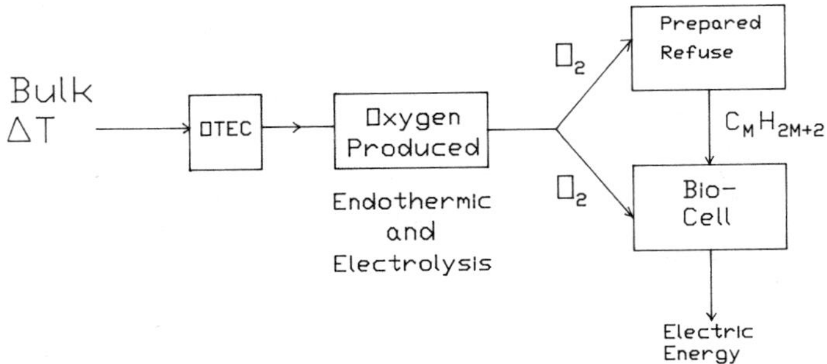

Figure 7.4 Interconnection of OTEC to bioconversion.

7.3 LINKAGE OF OTEC TO BIOELECTROCHEMICAL CELL SYSTEM

7.3.1 Introduction

Energy output from the OTEC plantship in the form of an oxygen-rich substance can enhance the conversion of prepared refuse to an ordered hydrocarbon C_mH_{2m+2} and also produce electric power output from the biocell. In both stages, the oxygen-rich material will act as the oxidizing agent to release the C_mH_{2m+2} hydrocarbon and as an input to the cathode of the biocell performing the oxidization of the ordered hydrocarbon into generating an electric potential output. The OTEC bioconversion system is shown in Fig. 7.4.

7.3.2 Generation of Ordered Hydrocarbon and Electric Energy[5]

Let G_{o_1} be the free enthalpy out from OTEC plant for an oxygen-rich agent; hence,

$$G_{o_1} = H_{o_1} + T_{o_1}S_{o_1} \tag{7.77}$$

The released ordered hydrocarbon C_mH_{2m+2} will possess a free enthalpy Gi_1 expressed by

$$G_{i_1} = H_{i_1} - T_oS_{i_1} + G_{i_c} \tag{7.78}$$

or from equation 3.33, we can write, taking into account change in chemical composition taking place in the biocell, the Gibbs function released as

[5] Reprinted, with permission from "The Conversion of Energy in Chemical Reactions" by Lothar Riekert, Energy Conversion, vol. 15, pp. 81–84, 1976 Pergamon Press, © 1976, Pergamon Press, Ltd.

$$G_{i_c} = \sum_i n_i \left(W^o_{i_c} + RT_o \ln \frac{M}{N} \right) \tag{7.79}$$

Also, from the preceding, we may identify the differential contribution of oxygen molar enthalpy from OTEC system as

$$[k_1 \Delta O_{DS} + k_2 \Delta O_{CS}] \text{ J/kg-mole}$$

Therefore,

$$G_{o_1} = k_1' \Delta O_{DS} + k_2' \Delta O_{CS} \tag{7.80}$$

where k_1' and k_2' are the fractional OTEC output of oxygen-rich continuum due to endothermic and conventional electrolysis of water, respectively.

Now if η_{i_1} is the efficiency of release of C_mH_{2m+2} from prepared refuse due to the oxidizing role of OTEC oxygen output, we can write that

$$\begin{aligned} G_{i_1} &= \eta i_1 [k_1' \Delta O_{DS} + k_2' \Delta O_{CS}] \\ &= \eta i_1 (H_{o_1} + T_{o_1} S_{o_1}) \end{aligned} \tag{7.81}$$

Turning next to the electrochemical reaction inside the biocell, where oxygen agent is injected at the cathode and synthetic fuel is fed at the anode, we can present an analytical argument similar to that of extracting ordered hydrocarbon, whereby G_{o_2} is the remaining free enthalpy of oxygen at the biocell cathode.

$$G_{o_2} = H_{o_t} + T_o S_o \tag{7.82}$$

$$= k_3' \Delta O_{DS} + k_4' \Delta O_{CS} \tag{7.83}$$

k_3', k_4' is the remaining fraction for the contribution of differential enthalpies for endothermic decomposition and conventional electrolysis of water for the release of oxygen.

Now if the biocell operational efficiency is designated as η_{bc}, we can say that

$$\eta_{bc} = \frac{en\Gamma}{(k_3' \Delta O_{DS} + k_4' \Delta O_{CS}) + \sum_i \mu_{iH} dn_{iH}} \tag{7.84}$$

where

Γ = Faraday's constant = 96.5×10^6 coulomb

$\sum_i \mu_{iH} dn_{iH}$ = the total free enthalpy of the ordered hydrocarbon injected at the anode of the biocell

n = the number of electron-moles associated with the reaction equation

Equation 7.14 can be rewritten in another form:

$$\eta_{bc} = \frac{eit}{k_3' \Delta O_{DS} \Delta_{ODS} + k_4' \Delta O_{CS} \Delta_{OCS} + \Delta h_H n_H} \tag{7.86}$$

where

Δh_H – $\sum_i \mu_{iH} d_{ni}$, which is known as the external change in total enthalpy

Δn_H – the molar concentration of the ordered hydrocarbon C_mH_{2m+2}

7.4 SUMMARY

In this chapter, system linkage of OTEC to several nodes for fuel release and generation of electric power were presented. Fuel release included hydrogen, oxygen, as well as ammonia, while power generating systems included the conventional-electromechanical hydrogen–oxygen–fuel cell, MHD power generator, and the bioelectrochemical cell.

Also, the concept of OTEC output storage and transport was shown by the coupling of a conventional plant powered by OTEC's huge ΔT to a redox flow cell, where electrical energy from OTEC could be stored in a chemical form and then transported to anyplace desired.

In the analysis presented, systematic expressions were outlined for the incremental enthalpies, molar concentrations, efficiencies, as well as modes of system linkages for the release of hydrogen, oxygen, and ammonia and for energy storage and transport.

The system linkages presented here were only some of the well-known modes of energy conversion; other useful and compatible systems could emerge in the near future, for example, solar energy, oceanic waves, vibrational energy, and nuclear plants.

7.5 SOLVED EXAMPLES

1. In the process of securing hydrogen and oxygen through endothermic decomposition and conventional electrolysis of water, energized by OTEC thermal output, express the combined product of external change in molar enthalpy and molar concentration for oxygen and hydrogen in terms of ammonia and ocean tropic water Reynolds numbers.

Solution Refer to Figure 7.1, where

$$\eta_2 = \frac{E_2}{E_1} \tag{1.a}$$

$$E_i = \text{proportional to } (R_w R_{NH_3})^{0.2345} \tag{1.b}$$

Therefore,

$$E_2 = a_1 \eta_2 (R_w R_{NH_3})^{0.2345} \tag{1.c}$$

where a_1 is a constant.

Also we see from equation 6.3

$$\eta_3 = \frac{\Delta O_{DS} n_{DS} + \Delta H_{DS} n_{DS}}{E_2} = \frac{\Delta_1 OH}{E_2} \tag{1.d}$$

Therefore,

$$\Delta_1 OH = a_1 \eta_2 \eta_3 (R_w R_{NH_3})^{0.2345} \tag{1.e}$$

where $\Delta_1 OH$ is the combined product of enthalpy external molar change and molar concentration due to endothermic decomposition of water toward separation of oxygen and hydrogen.

Next, we proceed to the process of electrolysis and separation into O_2 and H_2, also from Figure 7.1,

$$Q_1 = (1 - \eta_1)E_0 = \frac{E_1}{\eta_1}(1 - \eta_1) \tag{1.f}$$

$$Q_2 = (1 - \eta_2)E_1 = a_1(1 - \eta_2)(R_w R_{NH_3})^{0.2345} \tag{1.g}$$

Also

$$Q_1 = \frac{1 - \eta_1}{\eta_1} a_1 (R_w R_{NH_3})^{0.2345} \tag{1.h}$$

Therefore, from

$$Q_3 = Q_1 + Q_2 \tag{1.i}$$

$$= a_1 \left(\frac{1 - \eta_1}{\eta_1}\right)(R_w R_{NH_3})^{0.2345}$$

$$= a_1(1 - \eta_2)(R_w R_{NH_3})^{0.2345}$$

$$= a_1(R_w + R_{NH_3})^{0.2345}\left(\frac{1 - \eta_1}{\eta_1} + 1 - \eta_2\right) \tag{i.j}$$

Then,

$$\begin{aligned} Q_4 &= \eta_5 Q_3 \\ &= a_1 \eta_5 (R_w R_{NH_3})^{0.2345}\left(\frac{1 - \eta_1}{\eta_1} + 1 - \eta_2\right) \end{aligned} \tag{1.k}$$

From equation 6,

$$\Delta H_{CS} n_{HCS} + \Delta O_{CS} n_{OCS} = \Delta_2 OH \tag{1.l}$$

where

$$\Delta_2 OH = a_1 \eta_6 \eta_5 (R_w + R_{NH_3})^{0.2345}\left(\frac{1 - \eta_1}{\eta_1} + 1 - \eta_2\right) \tag{1.m}$$

Therefore, the combined product of external molar enthalpy change and molar concentration for O_2 and H_2 becomes, by adding equations 1.e and 1.m,

$$\Delta OH = \Delta_1 OH + \Delta_2 OH$$

$$= \left[a_1\eta_2\eta_3 + a_1\eta_5\eta_6 \left(\frac{1 - \eta_1}{\eta_1} + 1 - \eta_2 \right) \right] (R_w R_{NH_3})^{0.2345} \tag{1.n}$$

ΔOH could be linked to the predicted heat transfer coefficient U_{pred} according to equation 18,

$$\Delta OH = \frac{a_1}{0.217} U_{pred} \left[\eta_2\eta_3 + \eta_5\eta_6 \left(\frac{1 - \eta_1}{\eta_1} + 1 - \eta_2 \right) \right] \tag{1.p}$$

2. In MHD generator channel, two important parameters in connection with the magnetofluid interactions are the magnetic Reynolds number R_m and the interaction parameter I. Show the effect of OTEC's form plantship T on R_m and I.

Solution

$$R_m = \mu_o \sigma_o(T) V_o Y_o \tag{2.a}$$

and

$$I = \frac{\sigma_o(T) B_o^2}{P_o V_o} \tag{2.b}$$

where

μ_o = the reference magnetic permeability in henries per meter
V_o = the reference velocity of MHD fluid
Y_o = the reference dimension
P_o = the reference mass density in kilograms per cubic meter
$\sigma_o(T)$ = the reference electrical conductivity of MHD fluid or plasma

As discussed earlier the conducting plasma seeded with 1 percent potassium as an electrical conductivity $\sigma_o(T)$, follows three regions of separate linear relationship:

$$\sigma_1 = \tfrac{2}{15}T - 187, \quad \text{for } 1400 < t < 1700°\text{F} \tag{2.c}$$

$$\sigma_2 = \tfrac{9}{20}T - 725, \quad \text{for } 1700 < t < 2500°\text{F} \tag{2.d}$$

$$\sigma_3 = \tfrac{1}{25}T + 2000, \quad \text{for } 2500 < t < 5000°\text{F} \tag{2.e}$$

For an OTEC plantship of the foam type, energy transferred to the MHD channel per unit mass W is equal to $C_p(\Delta T)^2/2T$, where C_p is the specific heat of foam at constant pressure.

For $\Delta T = 20°\text{C}$,

$$T = (\tfrac{9}{51} \times 20 + 32) = 68°\text{F}$$

Therefore,

$$W_{OTEC} = C_p \frac{(68)^2}{2T} \tag{2.f}$$

or

$$T = \frac{1}{2} \frac{C_p(68)^2}{W_{OTEC}} = \frac{2312\, C_p}{W_{OTEC}} \ °\text{F} \tag{2.g}$$

Therefore, equations 2.c, 2.d, and 2.e become

$$\sigma_1 = \frac{2}{15}\left(\frac{2312C_p}{W_{\text{OTEC}}}\right) - 187 \tag{2.h}$$

$$\sigma_2 = \frac{9}{20}\left(\frac{2312C_p}{W_{\text{OTEC}}}\right) - 725 \tag{2.i}$$

$$\sigma_3 = \frac{1}{25}\left(\frac{2312C_p}{W_{\text{OTEC}}}\right) + 2000 \tag{2.j}$$

We can observe from equations 2.a and 2.b that R_m and I will experience regional change with respect to σ depending on the level of expanded work by OTEC foam system. Usually in highly conducting MHD plasma, perturbations in magnetofluid components proceed with the order of $(R_m I)^n$ product, and hence, thermal input from OTEC will have greater influence on the phenomenon of MHD interactions, where $n = 0, 1, 2, 3, \ldots$.

3. In the process of recharging the redox flow cell by subjecting a DC voltage across the catholyte-anolyte terminals, establish a correlation linking the activities or pressures of the cell reactants and their molar concentrations.

Solution From equations 7.67 and 7.68,

$$\Delta E_{\text{Fe}} = \Delta E_{\text{Ti}} \tag{3.a}$$

Therefore,

$$\left(\frac{P_{\text{Fe}^{+2}}}{P_{\bar{e}} P_{\text{Fe}^{+3}}}\right) T_{\text{Fe}} = \left(\frac{P_{\text{Ti}^{+4}} P_{\bar{e}}}{P_{\text{Ti}^{+3}}}\right) T_{\text{Ti}} \tag{3.b}$$

or

$$(P_{\bar{e}}) T_{\text{Fe}} = \left(\frac{P_{\text{Fe}^{+2}} P_{\text{Ti}^{+3}}}{P_{\text{Fe}^{+3}} P_{\text{Ti}^{+4}}}\right) T_{\text{Fe}} - T_{\text{Ti}} \tag{3.c}$$

where

T_{Ti} = the operating temperature of the titanium in the anolyte
T_{Fe} = the operating temperature of the ferromagnetic catholyte solution Equation 3.c could be written in another form:

$$P_{\bar{e}} = \frac{P_{\text{Fe}^{+3}} P_{\text{Ti}^{+4}}}{P_{\text{Fe}^{+2}} P_{\text{Ti}^{+3}}} \tag{3.d}$$

4. Equations 7.84 and 7.85 present two different but equivalent expressions for the theoretical efficiency of the biocell. Establish an emerging correlation among molar concentrations of oxygen, electrons, as well as generalized expanded work and molar enthalpies for the biocell.

Solution Equating 7.84 and 7.85 results in

$$en\Gamma[k_3'\Delta O_{DS} n_{ODS} + k_4'\Delta O_{CS} n_{OCS} + \Delta h_H n_H] - eit[k_3'\Delta O_{DS} + k_4'\Delta O_{CS} + \sum_i \mu_{iH} dn_{iH}] = 0 \tag{4.a}$$

Therefore,

$$(k_3'\Delta O_{DS} + k_4'\Delta O_{CS})[en\Gamma(n_{ODS} + n_{OCS}) - eit] = eit \sum_i \mu_{iH} dn_{iH} - en\Gamma\Delta h_H n_H = 0 \quad (4.b)$$

However,

$$eit \sum_i \mu_{iH} dn_{iH} = en\Gamma\Delta h_H n_H \quad (4.c)$$

Therefore, the emerging relationship from 4.b is

$$k_3'\Delta O_{DS} = -k_4'\Delta O_{CS} \quad (4.d)$$

$$n_{ODS} + n_{OCS} = \frac{it}{n\Gamma} \quad (4.e)$$

The significance of 4.d implies that $\Delta O_{DS}/\Delta O_{CS} = -k_4'/k_3'$ while equation 4.e signifies that the total molar concentration of oxygen out of endothermic decomposition of water plus that due to conventional electrolysis could be obtained accurately.

7.6 REVIEW QUESTIONS

1. Differentiate between the total enthalpy of a reactant and the molar formation enthalpy; then identify the molar concentration of reactants and products in the redox flow cell.
2. What is meant by the external change in molar enthalpy or total formation enthalpy? Under what condition is an external change in total enthalpy zero?
3. Express practical problems in securing pure H_2 and O_2 from endothermic and conventional electrolysis of H_2O.
4. Distinguish between pressure and activity of a reactant in an electrochemical reaction equation. Under what condition are they equal?
5. Discuss practical utilization of ammonia secured from OTEC plantship aside from recycling within OTEC.
6. Discuss the realization that extraction of H_2 and O_2 through the process of endothermic decomposition of water obeys in efficiency terms the Carnot cycle formula.
7. Outline the environmental impacts of the OTEC linkage to the hydrogen-oxygen fuel cell and storage battery system.
8. Referring to Figure 7.2, indicate the feasibility of having a conventional electromechanical power plant within a system interconnection involving OTEC, MHD plant, and the H_2-O_2 fuel cell. Show the best location for the electromechanical power plant.
9. Referring to Figure 7.3, indicate the best practical option from the sources of DC generator output, MHD generator output, or fuel-cell output to be applied on the redox flow cell for recharging.

10. Identify a practical problem of storing and transporting electric energy via ferric fluid. Why can't ferrous fluid be transferred to other load centers?

11. Discuss the concept of irreversibility defined by $ds\ dQ/T$ on the processes of recharge and discharge with respect to the redox flow cell.

$$ds = \text{the perturbation in entropy}$$
$$dQ = \text{the perturbation in heat}$$
$$T = \text{the absolute temperature}$$

12. Referring to Figure 7.4, identify the feasibility of the H_2-O_2 fuel cell within the system interconnection of OTEC and the biocell. Discuss the entire environmental impact of the new system.

7.7 PROBLEMS

1. Referring to Equation 7.18, express the molar concentration of H_2 out of endothermic and electrolysis processes followed by separation, $n_{H(C+D)S}$, in terms of external enthalpy changes. Modify the new relationship for irreversible performance indicating its implications of the overall reactions.
2. Modify equation 7.19 to solve for $n_{H(C+D)S}$, if $Q_1 = Q_2 = 0$, and the input for conventional electrolysis is from H_2-O_2 fuel cell powered by the output of endothermic decomposition of water.
3. Develop a solution for the molar concentration of produced NH_3 by compounding H_2 from the process of endothermic decomposition of water only with nitrogen from air liquefaction. Changes in enthalpies have to be in terms of imposed energy external changes. Structure a block diagram for the system interconnection to OTEC, with no conventional electrolysis.
4. Develop a systematic interconnection of an OTEC plantship for the direct production of ammonia and the recycling of ammonia back to the OTEC. Solve for the overall efficiency and the molar concentration of NH_3.
5. Refer to the result in Problem 4, considering that molar concentration of ammonia is a function of time and one-dimensional space variables. Develop a mathematical model in the complex frequency domain for the molar concentration of NH_3. Consider appropriate initial conditions.
6. Establish a system interconnection for the release of net H_2 and O_2 from OTEC plant with an expression for the overall efficiency and the molar concentration for H_2 and O_2 separately. Show under what conditions O_2 release is maximum and hence H_2 release is minimum. Consider that imposed enthalpy external change is variable.
7. Develop a systematic interconnection for the release of H_2 and O_2 from conventional electrolysis with energy input from OTEC plant. [Express the solution for the overall efficiency, molar concentration of O_2 and H_2 separately and the criterion of maximizing each one.]
8. Develop a systematic interconnection for OTEC plantship to an MHD and DC generator. Solve for the generator-produced emf, output power density, and efficiency. Express the loading factor defined as the ratio of the generator terminal voltage to the open-circuit voltage.

9. Using results secured from Problem 8, develop criterion for the MHD generator maximum current density, maximum power output, and maximum efficiency in terms of the MHD generator cylindrical geometrical dimensions, MGD plasma parameters, and OTEC's ΔT.

10. Repeat Problem 8 but for the interconnection of OTEC to power an MHD generating plant and the release of net O_2 and H_2.

11. Repeat Problem 9 but for the interconnection of OTEC to power an MHD generating plant and the maximum release of O_2 and H_2.

12. Consider the systematic interconnection of an OTEC plantship to a conventional fuel cell used to recharge a redox flow cell with the conversion of ferrous catholyte solution into ferric solution. Express the mathematical solution for the total system efficiency and give the criterion for its maximum limit.

13. Develop a systematic mathematical model for the interconnection of OTEC plantship to that of the redox flow cell. Refer to the redox cell dynamic model of Chapter 4. Consider time variation for ΔT, heat transfer coefficients, as well as water and ammonia Reynolds numbers. Consider the redox flow cell emf as the final output and OTEC's ΔT as its input.

14. From the unified model for OTEC linked to the redox flow cell established in Problem 13, express the criteria for the steady state as well as the initial limit of the cell output emf. Indicate the elements of system instability.

15. Establish a unified mathematical model for the linkage of OTEC plantship to the biocell, with ΔT as OTEC's input and the emf as the biocell output. Refer to the biocell model established earlier.

16. From the mathematical model of the OTEC-biocell system, develop criteria for the steady state and initial limit of the biocell emf. Indicate possible elements of instability.

17. Establish a mathematical model of energy system interconnection including redox flow cell, biochemical fuel cell, and H_2-O_2 fuel cell in the complex frequency domain with proper initial conditions. Then derive criteria for the initial system output emf and its steady-state limit. For the combined system, the input is fuel and the output is the final emf.

18. Repeat Problem 17 but for the interconnection of OTEC, conventional power plant, and MHD generating plant. Identify parametric instability in the total system.

19. Repeat Problem 18 but for the interconnection of dispersed conventional fuel-cell system, redox-flow-cell system, and MHD plant.

20. Repeat Problem 18 but for the interconnection of dispersed storage battery plant, OTEC-electromechanical plant, and redox-flow-cell system.

7.8 REFERENCES

1. Bogardus, E. H., D. A. Krueger, and D. A. Thompson. "Dynamic Magnetization in Ferrofluids." *Journal of Applied Physics,* Vol. 49 (1978), pp. 3422–3429.
2. Cichanski, W. J., and J. S. O'Connor. "Developmental Testing of a Concrete Cold Water Pipe for Ocean Thermal Energy Conversion Systems." *Proceed-*

ings of 3rd International Conference on Alternative Energy Sources, Miami, Florida, December 1980.

3. Colichman, E. L. "Preliminary Biochemical Fuel Cell Investigations." *Proceedings of the IEEE,* May 1963, pp. 812–819.
4. Denno, K. "Bio-Electromechanical Conversion of Refuse to Energy." *1979 Proceedings of the 25th Annual Technical Meeting of the Institute of Environmental Sciences,* pp. 316–321. Seattle, Washington, April 1979.
5. Denno, K. "Dynamic Modeling of Basic Types of Solid-State Power Inverters." *Proceedings of Midwest Power Symposium,* Manhattan, Kansas, October 1976.
6. Denno, K. "Dynamic Modeling of a Hydrogen Generating System Using Fusion Reactor Exhaust Plasma Interconnected to a Fuel Cell." *Proceedings of 6th Symposium on Engineering Problems of Fusion Research,* San Diego, California, November 1975.
7. Denno, K. "Equivalent Circuit Parametric Model of the Redox Flow Cell for Bulk Energy Storage." *Proceedings of the 3rd International Conference on Alternative Energy Sources,* Miami, Florida, December 1980.
8. Denno, K. "Generation Aspects of Weakened Fusion Plasma in MHD Channel." *Journal of Applied Science and Engineering,* 1978, pp. 213–224.
9. Denno, K. "Magnetic Transport Properties of Dipolar Conducting Liquids at Ambient Temperature." *Journal of Electrostatics,* 1979, pp. 345–360.
10. Denno, K. "Magnetohydrodynamic Perturbations for a Finite Magnetic Reynolds Number." *Journal of Applied Science and Engineering,* 1979, pp. 261–276.
11. Denno, K. "Problems in the Redox Flow Cell, Fuel Cell, Storage Battery and Harmonically Commutated Inverter." *Proceedings of the 1976 Canadian Communications Power Conference,* Montreal, Canada, November 1976.
12. Denno, K. "Simulating Criterion for Bio-Chemical Conversion of Refuse to Synthetic Fuel and Electric Power." *Proceedings of 3rd International Conference on Alternative Energy Sources,* Miami, Florida, December 1980.
13. Fish, J. D., and R. C. Axtman. "Utilization of Exhaust Energy for Fuel Production." *Proceedings of the 5th Symposium on Engineering Problems of Fusion Research,* Princeton, New Jersey, November 1973.
14. Pandolfini, P., G. L. Dugger, F. K. Hill, and W. H. Avery. "Alclad Aluminum, Folded-Tube Heat Exchanger for OTEC." *Proceedings of 3rd International Conference on Alternative Energy Sources,* Miami, Florida, December 1980.
15. Reikert, L. "The Conversion of Energy in Chemical Reactions." *Energy Conversion, an International Journal,* Vol. 15 (1975).
16. Richards, D. E., J. Francis, and G. L. Dugger. "Conceptual Designs for Commercial OTEC Ammonia Product Plantships." *Proceedings of 3rd International Conference on Alternative Energy Sources,* Miami, Florida, December 1980.
17. Sasscer, D., T. Tosteson, and T. Morgan. "Biofouling in Simulated OTEC Evaporator Tubes on AI Sea Research Platform at Punta Tuna." *Proceedings of 3rd International Conference on Alternative Energy Sources,* Miami, Florida, December 1980.
18. Sisler, F. D. "Electrical Energy from Biochemical Fuel Cells." *New Scientist,* Vol. 12 (1961).

19. Sutton, G. W., and A. Sherman. *Engineering Magnetohydrodynamics.* McGraw-Hill Book Co., New York, 1965.
20. Warshay, M., and L. O. Wright. "Cost and Size Estimates for an Electrochemical Bulk Energy Storage Concept." *NASA Technical Memorandum,* NASA TMX-3192, Lewis Research Center, Cleveland, Ohio, February 1985, 11 pages.
21. Thaler, Lawrence H. "Electrically Rechargeable Redox Flow Cells." *NASA Technical Memorandum,* NASA TMX-71549, Lewis Research Center, Cleveland, Ohio, August 1974.

Appendix A [1, 3, 4, 7]

Analytical Solution of the Power Coordination Equation

A.1 SOLUTION OF FIRST-ORDER PARTIAL DIFFERENTIAL EQUATION WITH VARIABLE COEFFICIENTS (CANONICAL SYSTEM)

The process of system optimization of interconnected energy sources ranging from static generating sources to electromechanical modes frequently involves solution of partial differential equations with variable coefficients.

It is the desire of the engineer and the researcher to look for a closed form of the solution instead of jumping into the process of numerical solution. The closed-form solution has direct reliance on the actual physical properties of the interconnected system and also mathematical testing for the physical realization of the system such as the concept of convergence, continuity, analyzability, compliance with boundary and initial conditions, and so on.

In this appendix, the process of partial linearization based on converting the partial differential equation to a set of canonical system is ultimately presented, coupled with a practical example.

The expression

$$G(x, y, u, u_x, u_y) = 0 \tag{A.1.1}$$

is a first-order partial differential equation, where

$$u = f(x, y) \tag{A.1.2}$$

$$u_x = \frac{\partial u}{\partial y} \tag{A.1.3}$$

$$u_y = \frac{\partial u}{\partial y} \tag{A.1.4}$$

or

$$a(x, y, u)u_x + b(x, y, u)u_y = C[x, y, u] \tag{A.1.5}$$

u is an integral of equation A.1.5 and S_u, an integral surface of u. The S_u of a surface is defined by:

$$[x, y, z = u(X, y)]$$

Also, we can define

$$\begin{aligned} N_u(x, y, z) &= \text{a normal vector to the integral surface } S_u \\ \overline{K}(x, y, z) &= a(x, y, z), b(x, y, z), c(x, y, z) \end{aligned} \tag{A.1.6}$$

Therefore,

$$\begin{aligned} \overline{K} \cdot \overline{N}_n &= aU_x + bU_y - c \\ &= 0, \end{aligned} \tag{A.1.7}$$

if U is an integral of A.1.5. Hence, $\overline{K}(x, y, z)$ is called the characteristic direction vector at (x, y, z). Then, we can say that S_u is an integral surface if at each of its points the characteristic direction is tangent to S_u. By "characteristic" we mean space curves whose tangents coincide with the characteristic direction at each of their points. Therefore, we can set

$$\frac{dx}{a(x, y, z)} = \frac{dy}{b(x, y, z)} = \frac{dz}{c(x, y, z)} = ds \tag{A.1.8}$$

or

$$\begin{aligned} \frac{dx}{ds} &= a(x, y, z) \\ \frac{dy}{ds} &= b(x, y, z) \\ \frac{dz}{ds} &= c(x, y, z) \end{aligned} \tag{A.1.9}$$

where

$$\begin{aligned} x &= \alpha(x) \\ y &= \beta(s) \\ z &= \gamma(s) \end{aligned} \tag{A.1.10}$$

Through each point (x_o, y_o, z_o) is the passage of one characteristic, where

$$x = x(x_o, y_o, z_o)$$
$$y = y(x_o, y_o, z_o) \tag{A.1.11}$$
$$z = z(z_o, y_o, z_o)$$

Hence, a surface S_u that could be generated by one family of characteristics is an integral surface.

Now, as we mentioned earlier, let S_u be an integral surface, and let (x_o, y_o, z_o) be a point of this surface. Then, consider the solution

$$x = x(s)$$
$$y = y(s) \tag{A.1.12}$$

Therefore,

$$\frac{dx}{ds} = a[x, y, u(x, y)]$$
$$\frac{dy}{ds} = b[x, y, u(x, y)] \tag{A.1.13}$$

and

$$x(0) = x_o$$
$$y(0) = y_o \tag{A.1.14}$$

Then for the curve,

$$x = x(s)$$
$$y = y(s) \tag{A.1.15}$$
$$z = u[x, y(s)]$$

We shall have,

$$\frac{dz}{ds} = U_x \frac{dx}{ds} + U_y \frac{dy}{ds}$$
$$= aU_x + bU_y \tag{A.1.16}$$
$$= c(x, y, u)$$

Therefore, we conclude such a curve is a characteristic.

The Concept of Monge Cone (7.1)

Monge cone is the envelope of all tangent planes to the surface S_u; for each p and q, there is a tangent plane called planer element, where

$$p = \frac{\partial u}{\partial x}, \qquad q = \frac{\partial u}{\partial y} \tag{A.1.17}$$

Consider, again,

$$G(x, y, z, p, q) = 0 \tag{A.1.18}$$

We can say $(p, q, -1)$ is normal to the surface $z(x, y)$. At every point on the surface z, the tangent plane to the surface and the tangent plane to the monge cone coincide. Therefore,

$$dz = p\,dx + q\,dy \tag{A.1.19}$$

The family of equations of the tangent surfaces for equation A.1.18 is expressed below without derivation:

$$\frac{dx}{dy} = \frac{dy}{fq} = \frac{dz}{pfp + qfp} = \frac{-dP}{fx + pfy} = \frac{-dq}{fy + qfz} = ds \tag{A.1.20}$$

The set in equation A.1.20 in effect represents partial linearization of A.1.17, where

$$fp = \frac{\partial F}{\partial P}, \qquad fq = \frac{\partial F}{\partial q}, \qquad fz = \frac{\partial F}{\partial Z} \tag{A.1.21}$$

Example

$$p^2 + qy = z = f \tag{A.1.a}$$

Using the set of A.1.20 with respect to A.1.a, we obtain

$$\frac{dx}{2p} = \frac{dy}{y} = \frac{dz}{2p^2 + qy} = -\frac{dP}{P} = -\frac{dq}{0} \tag{A.1.b}$$

Therefore,

$$q = a \tag{A.1.c}$$

$$P = \pm\sqrt{z - ay} \tag{A.1.d}$$

$$dz = \pm\sqrt{x - ay}\,dx + ady \tag{A.1.e}$$

Therefore,

$$\frac{dz - ady}{\sqrt{z - ay}} = \pm dx \tag{A.1.f}$$

$$\frac{dz}{\sqrt{z - ay}} - \frac{ady}{\sqrt{z - ay}} = \pm dx \tag{A.1.g}$$

Therefore,

$$2(z - ay) = \pm(x + b)^2 \tag{A.1.h}$$

or

$$4(z - ay) = (x + b)^2 \tag{A.1.i}$$

Equation A.1.i is the complete solution.

A.2 POWER SYSTEM SYNTHESIS FROM SOLUTION OF OPTIMUM TRANSMISSION LOSS COEFFICIENTS[1]

Abstract

This paper presents an analytical approach to the problem of expressing the transmission loss (B) matrix in terms of optimum scheduling of generating sources and their fuel-cost data. Conversion of compatible matrix differential equations to a canonical system resulted in a closed-form solution for the B matrix. The result obtained can be applied in selecting the optimum power system, predicting bus phase angles, and updating the B matrix at any loading.

Introduction

Determination of the optimum system will depend on economic evaluation, reliability and stability problems.[1,4] The generating sources of the system are primarily electrochemical devices comprising fuel cells and storage devices that are interconnected with electromechanical generating sources.

The primary goal of the investigation centers on various aspects of the steady-state and dynamic system problems for the electrochemical system and then with its interconnection with the electromechanical system.

However, as a basic first step along the path of this investigation that may serve in the following stages, a pure theoretical analysis is conducted toward the establishment of the best system from among a large number of alternative plans, each with a different number of generating sources, generating capacities, and locations in the system.

Hence, in the search for the best economically balanced power system where transmission losses are significant, the coordination equation between incremental production cost and incremental transmission loss represents an accurate criterion once the electric system configuration has been established on other grounds, upon which an optimum scheduling of generation

[1] © 1973 IEEE. Reprinted, with permission, from IEEE Transactions on Power Apparatus and Systems, Vol. PAS-92, No. 6, November/December 1973. Paper entitled: "Power System Synthesis From Solution of Optimum Transmission Loss Coefficients" by K. Denno, Newark College of Engineering, Newark, New Jersey.

can be obtained. However, prior to performance of an iteration process, transmission coefficients must be calculated by (1) the simple B parameters formula, which is based on the fact that each load bus remains a constant fraction with respect to total loads; (2) the generalized loss formula, which is based on the assumption that the load bus is composed of two components, one of which is constant and the other a linear function of the total load; or (3) resorting to the phase angle method of calculation. The foregoing methods of calculating transmission coefficients consider, in most cases, one base loading condition that will be used in all other loadings.

The fact is, those interacting transmission loss parameters or coefficients are functions of changes in load pattern, voltage magnitude, and phase angle, in addition to changes in the generating sources and their locations in the power system.

In the process of comparison among several optimum power systems, the possibility exists for varieties of options with respect to the nature and type of fuel, output of each source within its range or capacity and locations, and different types of generating sources. All these will reflect the fact that there will be a large spectrum of optimum power systems, each possessing a quite distinct matrix for transmission coefficients.

Accordingly, each optimum power system comprising fossil-fuel generators, fuel cells, storage batteries, and nuclear generators or combinations of those sources will have its own distinct transmission matrix elements subject to the output of each source within its capacity, location, total generation, voltage magnitude, phase angle, and load pattern.

Statement of the Problem

Consider several alternatives of electric power system configurations (as shown in Fig. A.2.1), each satisfying an optimum state based on the coordination and constraint equations A.2.1 and A.2.2, respectively. Each system is characterized by the following features:

1. Locations of the generating sources
2. Output of each source within its range or capacity
3. Fuel cost and rate of consumption
4. Mode of generating system and its interacting parameters (e.g., centralized, dispersed, or combination)
5. Voltage phase angle and magnitude at each bus

The coordination equation is[2]

$$\frac{\partial F_i}{\partial P_i} + \lambda \frac{\partial W}{\partial P_i} = \lambda \tag{A.2.1}$$

[2] See Kirchmayer and Stagg in Endnotes.

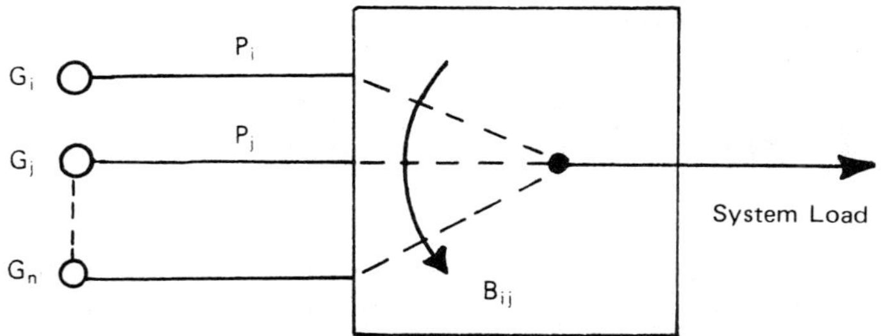

Figure A.2.1 Schematic representation for one of several alternative power systems that can be established on different grounds

subject to the following constraint equation

$$\phi(P_1, P_2, \ldots, P_n) = \sum_n P_n - P_r - P_L = 0 \qquad \text{(A.2.2)}$$

$$W \cong \sum_i \sum_j P_i B_{ij} P_j \qquad \text{(A.2.3)}$$

assuming that the load currents vary together as a constant complex fraction of the total load current.

λ = cost of received power in dollars per megawatt hour
W = total transmission loss
P_r = given received load
P_i = capacity of plant i in megawatts
F_i = fuel cost of plant i in dollars per hour

Therefore, each alternative optimum system is characterized by corresponding values of transmission line coefficients.

Hence, solution of the B coefficients will identify the structure of such a system, taking into account the following assumptions:

1. The system contains n sources of fossil, electrochemical, and nuclear fuel, each characterized by a specific fuel-cost curve.
2. The transmission coefficient or parameter B_{ij} is related to and depends on P_j and P_i as indicated:

$$P_j = \phi(P_i) \qquad \text{(A.2.4)}$$

$$B_{ij} = \psi(P_i) \qquad \text{(A.2.5)}$$

Therefore, B_{ij} is an implicit function of P_j. Equations A.2.4 and A.2.5 imply general recognition that B_{ij} is strongly connected to transfers among generating sources.

3. Since B_{ij} is assumed to be a general function of sources, the total transmission loss W is assumed to take the simple form:

$$W = \sum_i \sum_j P_i B_{ij} P_j \tag{A.2.6}$$

$$= F(P_i, B_{ij}, P_j) \tag{A.2.7}$$

The problem centers on obtaining a general solution of B_{ij} from the coordination equation A.2.1 in terms of source output within its range or capacity and their economic cost data. This implies taking economic conditions, reliability, and stability into account in the derivation.

Solution of Transmission Loss Coefficients Matrix[3,4,5,7]

Rewrite equation A.2.1:

$$\frac{\partial F_i}{\partial P_i} + \lambda \frac{\partial W}{\partial P_i} = \lambda$$

substituting from equation A.2.6 and assuming that the general incremental fuel-cost curve for varieties of generating sources such as conventional generators, fuel cells, storage batteries, and nuclear is linear, such that

$$\frac{\partial F_i}{\partial P_i} = F_{ii} P_i + f_i \tag{A.2.8}$$

F_{ii} = slope of incremental fuel-cost curve
f_{ii} = incremental cost of plant i at zero output

Therefore,

$$F_{ii} P_i + f_i + \lambda \left(B_{ij} P_j + \frac{\partial W}{\partial P_j} \frac{dP_j}{dP_i} + \frac{\partial W}{\partial B_{ij}} \frac{dB_{ij}}{dP_i} \right) = \lambda \tag{A.2.9}$$

assuming that tie flows interconnecting a given area with other areas are independent of the generation allocation within the area under study.

In the foregoing equations, the Einstein summation convention is adopted where the occurrence in a product of the same letter carrying the same index represents a summation over that index. Equation A.2.9 can be written as

$$F_{ii} P_i + f_i + \lambda \left(B_{ij} P_j + P_i B_{ij} \frac{dP_j}{dP_i} + P_i P_j \frac{dB_{ij}}{dP_i} \right) = \lambda \tag{A.2.10}$$

Note that P_i, B_{ij}, and P_j are symmetrical matrices. Equation A.2.1 can also be written in another form:

$$\frac{\partial F_j}{\partial P_j} + \lambda \frac{\partial W}{\partial P_j} = \lambda \tag{A.2.11}$$

Then equation A.2.10 can be expressed in a compatible form as

$$F_{jj}P_j + f_j + \lambda\left(P_iB_{ij} + \frac{\partial W}{\partial P_i}\frac{dP_i}{dP_j} + \frac{\partial W}{\partial B_{ij}}\frac{dB_{ij}}{dB_j}\right) = \lambda \qquad \text{(A.2.12)}$$

$$F_{jj}P_j + f_j + \lambda\left(P_iB_{ij} + B_{ij}P_j\frac{dP_i}{dP_j} + P_iP_j\frac{dB_{ij}}{dP_j}\right) = \lambda \qquad \text{(A.2.13)}$$

Equations A.2.10 and A.2.13 are two compatible matrix differential equations. Solving simultaneously for the two quantities dE_{ij}/dP_j and dB_{ij}/dP_i will give

$$\frac{dB_{ij}}{dP_j} = \frac{4\lambda^2P_iB_{ij}^2P_j}{2\lambda P_iP_j(P_{ii}P_i + f_i - \lambda)} - (F_{ii}P_i + f_i - \lambda + 2\lambda B_{ij}P_j) \cdot \left[\frac{P_{jj}P_j + f_j - \lambda + 2\lambda P_iB_{ij}}{2\lambda P_iP_j(F_{ii}P_i + f_i - \lambda)}\right] \qquad \text{(A.2.14)}$$

$$\frac{dB_{ij}}{dP_i} = \frac{4\lambda^2P_iB_{ii}^2P_j}{2\lambda P_iP_j(P_{jj}P_j + f_j - \lambda)} - (F_{ii}P_i + f_i - \lambda + 2\lambda B_{ij}P_j) \cdot \left[\frac{P_{jj}P_j + f_j - \lambda + 2\lambda P_iB_{ij}}{2\lambda P_iP_j(F_{jj}P_j + f_j - \lambda)}\right] \qquad \text{(A.2.15)}$$

The intention now is to obtain a solution for B_{ij} in terms of P_i, P_j and their cost data. The differential equation A.2.10 can be written in general form:

$$F\left(P_i, P_j, B_{ij}, \frac{\partial B_{ij}}{\partial P_i}, \frac{\partial B_{ij}}{\partial P_j}\right) = 0 \qquad \text{(A.2.16)}$$

Also, similarly, equation A.2.13 can be expressed in general form:

$$G\left(P_i, P_j, B_{ij}, \frac{\partial B_{ij}}{\partial P_i}, \frac{\partial B_{ij}}{\partial P_j}\right) = 0 \qquad \text{(A.2.17)}$$

And, hence,

$$F - G = \psi = 0 \qquad \text{(A.2.18)}$$

represents another compatible differential equation that can be solved for P_i, P_j, B_{ij}, and even for $\partial B_{ij}/\partial P_i$ and $\partial B_{ij}/\partial P_j$ by conversion to a canonical system, expressed below [7]:

$$\frac{dP_i}{d\psi_p} = \frac{dP_j}{d\psi_q} = \frac{-dQ}{\psi_{p_j} + Q\psi_{B_{ij}}} = \frac{-dP}{\psi_{p_i} + P\psi_{B_{ij}}} = \frac{dB_{ij}}{P\psi_p + Q\psi_q} \qquad \text{(A.2.19)}$$

where

$$P = \frac{\partial B_{ij}}{\partial P_i} \qquad Q = \frac{\partial B_{ij}}{\partial P_j}$$

$$\psi_{p_j} = \frac{\partial \psi}{\partial P_j} \qquad \psi_{B_{ij}} = \frac{\partial \psi}{\partial B_{ij}}$$

Apply A.2.18 on A.2.19 to obtain

$$\frac{dP}{F_{ii}Q} = -\frac{dQ}{F_{jj}P} \tag{A.2.20}$$

$$F_{jj}P^2 + F_{ii}Q^2 = C^2 \tag{A.2.21}$$

P and Q are those expressed in equations A.2.14 and A.2.15, respectively. C^2 is a constant, which was found by substituting A.2.14 and A.2.15 into A.2.21, then using these boundary conditions that are deduced from equation A.2.1:

$$B_{ij} = 0$$

for

$$P_i = \frac{\lambda - f_i}{F_{ii}} \tag{A.2.22}$$

and

$$P_j = \frac{\lambda - f_j}{F_{jj}} \tag{A.2.23}$$

The process gave $C^2 = 0$, and equation A.2.19 becomes

$$F_{jj}\left(\frac{\partial B_{ij}}{\partial P_i}\right)^2 + F_{ii}\left(\frac{\partial B_{ij}}{\partial P_j}\right)^2 = 0 \tag{A.2.24}$$

substituting A.2.14 and A.2.15 into A.2.24 and solving for B_{ij} will yield the result

$$\begin{aligned} B_{ij} = {} & [-F_{ii}F_{jj}P_iP_j + (\lambda F_{ii} - F_{ii}f_j)P_i \\ & + (\lambda F_{jj}F_{jj}f_i)P_j + (\lambda f_i + \lambda f_j - f_if_j - \lambda^2)]/ \\ & [2\lambda F_{ii}P_i^2 + 2\lambda F_{jj}P_j^2 + (2\lambda f_i - 2\lambda^2)P_i \\ & + (2\lambda f_j - 2\lambda^2)P_j] \end{aligned} \tag{A.2.25}$$

Equation A.2.25 expresses the B constants for a power system of minimum dollars input for a given received load in terms of optimum scheduling of generation and fuel-cost data. However, other physically realizable degrees of freedom in the calculation of the B matrix can be added such as a total minimum transmission loss and a specified individual plant output within its range or capacity.

Conclusions

The solution of transmission line interaction parameters or coefficients expressed by equation A.2.25 enables synthesizing a corresponding optimum power system from a given group of electric system configurations established on different grounds with respect to location of generating sources, characteristics of buses, pattern of loads, specified total generation, specified plant capacity, and minimum total transmission losses.

1. Specified total generation and individual plant output will be represented by a corresponding optimum value of B coefficients matrix. Such matrix or group of matrices can serve as bases for synthesizing the optimum power system required for every constraint assumed. These constraints could be any one or more of the following:
   a. Specified total generation
   b. Specified individual plant capacity
   c. Specified total load received
   d. Specified minimum total transmission losses.

Example

For the following eight plant systems with their fuel-cost data indicated, as in Appendix I, the B_{ij} matrix elements were calculated according to specified total generation and specified individual plant megawatt capacities. The constraint on the value of λ is that corresponding to minimum total transmission losses for each case. In Appendix II, three matrices for B coefficients are listed each for specified total generation and individual plant capacity under the constraint of minimum total transmission losses for the value of λ. The values for the B matrices were computerized, and a typical flow diagram is shown in Figure A.2.2.

2. An optimum value of the B matrix can serve as a reliable tool in the process of checking economic scheduling of generating sources under certain constraints, as in the coordination between incremental production costs and incremental transmission losses. In each iteration process recalculation of the B elements will affirm the fact of the dependence of the transmission loss coefficients on the generating source outputs, their location, and their load pattern.

Example [4]

In a two-plant system having the fuel-cost data given in Figure A.2.3, and with peak total generation of 200 MW, an economic scheduling of gener-

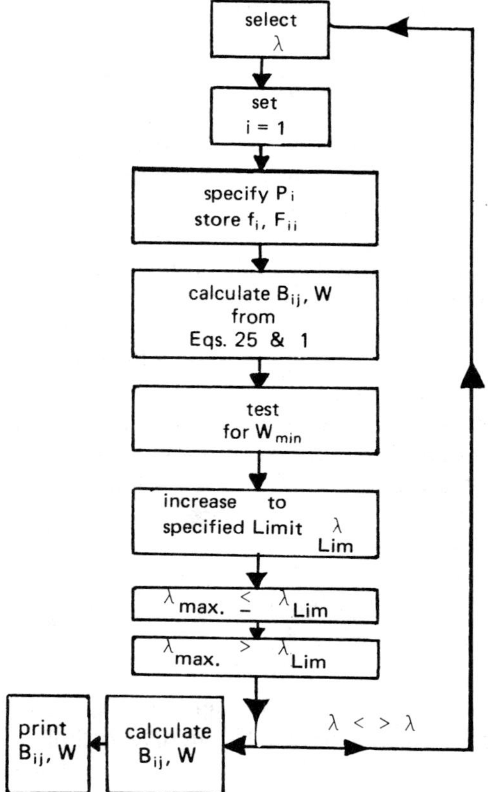

Figure A.2.2 Flow chart of computer program for optimum B matrix of minimum total transmission loss system.

ation for a specified total generation has been calculated by equation A.2.25 and checked with the coordination equation A.2.1. In Figure A.2.3 are shown the two curves for optimum generation schedule of the two plants.

3. The author sees this additional possibility, that from the principles of power system synthesis, knowledge of optimum values of the elements for the B matrix can be used in some special case to calculate the phase angles of any bus, given a specified ratio of real to reactive powers, which is related to the reverse procedure of calculating B constants from the phase angles.

Example

In cases where small changes in bus phase angles are assumed such that

$$B_{ij} = K_{ij}R_{ij} \tag{A.2.26}$$

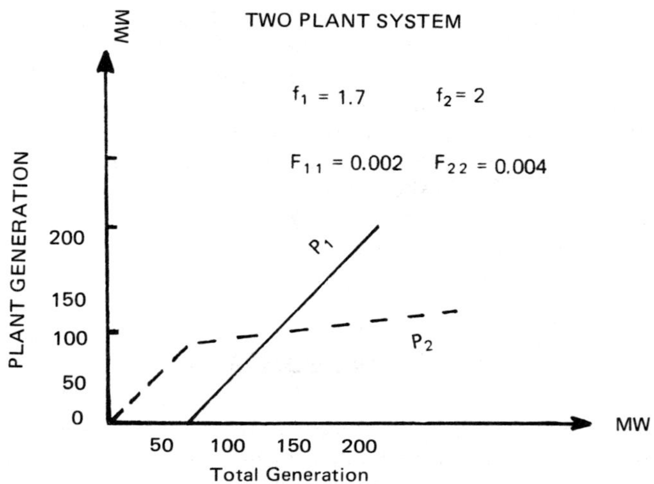

Figure A.2.3 Scheduling of generation.

where R_{ij} is the symmetric resistances of the system in endnote 3

$$K_{ij} = \frac{1}{|V_i||V_j|} [(1 + S_i S_j) \cos \phi_{ij} + (S_i - S_j) \sin \phi_{ij}] \tag{A.2.27}$$

and S_i is the ratio of real power to reactive power at bus i; therefore, if R_{ij} is known, reliable information about ϕ_{ij} can be obtained.

4. In the process of iteration to obtain economic scheduling of generation, recalculation of the transmission loss coefficients can be done without resort to another round of load flow calculations. Recalculation of the B matrixes at each specified scheduling of generation is consistent with the changing pattern of those matrices at each generation schedule, provided that Δp's and $\Delta\lambda$'s are known that make the successive λ's those that satisfy the coordination equation A.2.1.

ENDNOTES

1. Brownlee, W. R. "Coordination of Incremental Fuel Costs and Incremental Transmission Losses by Functions of Voltage Phase Angles." *AIEE Transactions on Power System and Apparatus,* 1954, pp. 529–541.
2. Elgerd, O. I. *Electric Energy Systems Theory*. New York: McGraw-Hill Book Co., 1971.

3. Kron, G. "Tensorial Analysis of Integrated Transmission Systems." *AIEE Transactions on Power System and Apparatus,* Vol. 70, (1951), pp. 1239–1246.
4. Kirchmayer, L. K., and G. W. Stagg. "Evaluation of Methods of Coordinating Incremental Fuel Costs and Incremental Transmission Losses." *AIEE Transactions on Power System and Apparatus,* January 1952, pp. 513–521.
5. Mickle, M. H., and T. W. Sze. *Optimization in Systems Engineering.* Scranton, PA: Intext Educational Publishers, 1972.
6. Stagg, G. W., and A. H. El-Abiad. *Computer Methods in Power System.* New York: McGraw-Hill Book Co., 1968.
7. Garabedian, P. R. *Partial Differential Equations.* New York: John Wiley & Sons, 1964.

ACKNOWLEDGMENT

The author wishes to thank the Middle Atlantic Power Research Committee for their support in sponsoring the research work presented in this paper. The author would also like to acknowledge the valuable comments of all member companies made with respect to the paper and in particular Mr. W. S. Ku of Public Service Electric and Gas Company and Mr. J. H. Rumbaugh of Potomac Electric Power Company.

APPENDIX I

TABLE I Fuel-cost Data of Eight Plant Systems

Plant	F_{ii}	f_i
1	0.0082	1.28
2	0.0044	0.795
3	0.0019	1.809
4	0.00429	0.657
5	0.00222	0.889
6	0.012	0.300
7	0.0208	0.635
8	0.0127	0.572

TABLE II Specified Scheduling of Generation

P_1	P_2	P_3	P_4	P_5	P_6	P_7	P_8
Total Peak Generation = 2,000 MW							
260	210	270	420	310	240	110	180
67% Total Generation							
145	210	75	285	310	150	75	85
41% Total Generation							
45	210	100	160	310	80	45	60

APPENDIX II

TABLE I

B_{ij} Matrix

Total Peak Generation $\lambda = 2.9$, $W_{min} = 17$ MW

$.2567 \times 10^{-3}$	$-.3597 \times 10^{-3}$	$-.1228 \times 10^{-3}$	$-.1131 \times 10^{-3}$	$-.278 \times 10^{-3}$	$.3305 \times 10^{-3}$	$.3157 \times 10^{-4}$	$.1358 \times 10^{-4}$
$-.3597 \times 10^{-3}$	$.5808 \times 10^{-3}$	$.2147 \times 10^{-3}$	$.1854 \times 10^{-3}$	$.4901 \times 10^{-3}$	$.3304 \times 10^{-3}$	$.3736 \times 10^{-4}$	$-.1704 \times 10^{-4}$
$-.1228 \times 10^{-3}$	$.2147 \times 10^{-3}$	$.8358 \times 10^{-4}$	$.6923 \times 10^{-4}$	$.1919 \times 10^{-3}$	$-.9868 \times 10^{-4}$	$-.1184 \times 10^{-4}$	$-.5539 \times 10^{-5}$
$-.1131 \times 10^{-3}$	$.1854 \times 10^{-3}$	$.6923 \times 10^{-4}$	$.5933 \times 10^{-4}$	$.1581 \times 10^{-3}$	$-.1009 \times 10^{-3}$	$-.1156 \times 10^{-4}$	$-.5304 \times 10^{-5}$
$-.2786 \times 10^{-3}$	$.4901 \times 10^{-3}$	$.1919 \times 10^{-3}$	$.1581 \times 10^{-3}$	$.4404 \times 10^{-3}$	$-.2215 \times 10^{-3}$	$-.2669 \times 10^{-4}$	$-.1250 \times 10^{-4}$
$.3305 \times 10^{-3}$	$-.3304 \times 10^{-3}$	$-.9868 \times 10^{-4}$	$-.1009 \times 10^{-3}$	$-.2215 \times 10^{-3}$	$-.19 \times 10^{-2}$	$.9247 \times 10^{-4}$	$.271 \times 10^{-4}$
$.3157 \times 10^{-4}$	$-.3736 \times 10^{-4}$	$-.1184 \times 10^{-4}$	$-.1156 \times 10^{-4}$	$-.2669 \times 10^{-4}$	$.9248 \times 10^{-4}$	$.5222 \times 10^{-5}$	$.1993 \times 10^{-5}$
$.1358 \times 10^{-4}$	$-.1704 \times 10^{-4}$	$-.5534 \times 10^{-5}$	$-.5304 \times 10^{-5}$	$-.125 \times 10^{-4}$	$.271 \times 10^{-4}$	$.1993 \times 10^{-5}$	$.8031 \times 10^{-6}$

TABLE II

B_{ij} Matrix

67% Total Peak Generation $\lambda = 2$, $W_{min} = 19.5$ MW

$.4545 \times 10^{-3}$	$-.1894 \times 10^{-3}$	$-.4567 \times 10^{-4}$	$-.8788 \times 10^{-4}$	$-.1863 \times 10^{-3}$	$.2464 \times 10^{-3}$	$.2017 \times 10^{-3}$	$.7604 \times 10^{-4}$
$-.1894 \times 10^{-3}$	$.8772 \times 10^{-4}$	$.1953 \times 10^{-4}$	$.4059 \times 10^{-4}$	$.9414 \times 10^{-4}$	$-.8431 \times 10^{-4}$	$-.7172 \times 10^{-4}$	$-.2866 \times 10^{-4}$
$-.4567 \times 10^{-4}$	$.1953 \times 10^{-4}$	$.4619 \times 10^{-5}$	$.9058 \times 10^{-5}$	$.196 \times 10^{-4}$	$-.2345 \times 10^{-4}$	$-.1941 \times 10^{-4}$	$-.7439 \times 10^{-5}$
$-.8788 \times 10^{-4}$	$.4059 \times 10^{-4}$	$.9059 \times 10^{-5}$	$.1879 \times 10^{-4}$	$.4346 \times 10^{-4}$	$-.3929 \times 10^{-4}$	$-.3339 \times 10^{-4}$	$-.1333 \times 10^{-4}$
$-.1863 \times 10^{-3}$	$.9414 \times 10^{-4}$	$.1960 \times 10^{-4}$	$.4346 \times 10^{-4}$	$.11 \times 10^{-3}$	$-.7317 \times 10^{-4}$	$-.6361 \times 10^{-4}$	$-.2634 \times 10^{-4}$
$.2464 \times 10^{-3}$	$-.8432 \times 10^{-4}$	$-.2345 \times 10^{-4}$	$-.3929 \times 10^{-4}$	$-.7318 \times 10^{-4}$	$.254 \times 10^{-3}$	$.1744^{3} \times 10^{-3}$	$.5341 \times 10^{-4}$
$.2017 \times 10^{-3}$	$-.7173 \times 10^{-4}$	$-.1941 \times 10^{-4}$	$-.334 \times 10^{-4}$	$-.6361 \times 10^{-4}$	$.1744 \times 10^{-3}$	$.1267 \times 10^{-3}$	$.4113 \times 10^{-4}$
$.7604 \times 10^{-4}$	$-.2866 \times 10^{-4}$	$-.7439 \times 10^{-5}$	$-.1333 \times 10^{-4}$	$-.2634 \times 10^{-4}$	$.5341 \times 10^{-5}$	$.4112 \times 10^{-4}$	$.1428 \times 10^{-4}$

TABLE III

B_{ij} Matrix

41% Total Peak Generation $\lambda = 1.6$, $W_{min} = 18.48$ MW

$.9603 \times 10^{-5}$	$.1554 \times 10^{-4}$	$-.5303 \times 10^{-5}$	$-.3277 \times 10^{-4}$	$-.173 \times 10^{-5}$	$-.8304 \times 10^{-4}$	$-.4662 \times 10^{-4}$	$-.6226 \times 10^{-4}$
$-.1554 \times 10^{-4}$	$.3138 \times 10^{-4}$	$-.5766 \times 10^{-4}$	$-.666 \times 10^{-4}$	$-.4006 \times 10^{-5}$	$-.1321 \times 10^{-3}$	$-.6689 \times 10^{-4}$	$-.9883 \times 10^{-4}$
$-.5303 \times 10^{-5}$	$-.5766 \times 10^{-4}$	$-.1104 \times 10^{-5}$	$.1445 \times 10^{-3}$	$-.3229 \times 10^{-5}$	$.4278 \times 10^{-4}$	$.166 \times 10^{-4}$	$.317 \times 10^{-4}$
$-.3277 \times 10^{-4}$	$-.666 \times 10^{-4}$	$.1445 \times 10^{-3}$	$.1414 \times 10^{-3}$	$.8554 \times 10^{-5}$	$.2785 \times 10^{-3}$	$.1405 \times 10^{-3}$	$.2082 \times 10^{-3}$
$-.173 \times 10^{-5}$	$-.4006 \times 10^{-5}$	$-.3229 \times 10^{-5}$	$.8554 \times 10^{-5}$	$.576 \times 10^{-6}$	$.1458 \times 10^{-4}$	$.7025 \times 10^{-5}$	$.1008 \times 10^{-4}$
$-.8304 \times 10^{-4}$	$-.1322 \times 10^{-3}$	$.4279 \times 10^{-4}$	$.2785 \times 10^{-3}$	$.1458 \times 10^{-4}$	$.719 \times 10^{-3}$	$.4075 \times 10^{-3}$	$.5392 \times 10^{-3}$
$-.4662 \times 10^{-4}$	$-.6689 \times 10^{-4}$	$.166 \times 10^{-4}$	$.1405 \times 10^{-3}$	$.7025 \times 10^{-5}$	$.4075 \times 10^{-3}$	$.2467 \times 10^{-3}$	$.3061 \times 10^{-3}$
$.6226 \times 10^{-4}$	$-.9883 \times 10^{-4}$	$-.3170 \times 10^{-4}$	$.2082 \times 10^{-3}$	$.1088 \times 10^{-3}$	$.5392 \times 10^{-3}$	$.3061 \times 10^{-3}$	$.4044 \times 10^{-3}$

Appendix B [1, 4, 6, 7]

Laplace Transform of Selected Functions

$f(t)$	$F(s)$
1. $U_0(t)$ unit impulse	1
2. $U_{-1}(t)$ unit step	$\frac{1}{S}$
3. $U_{-2}(t)$	$\frac{1}{S^2}$
4. e^{-at}	$\frac{1}{S+a}$, $R_e(S+a)$ 0
5. $\sin \beta t$	$\frac{\beta}{S^2+\beta^2}$
6. $\cos \beta t$	$\frac{S}{S^2+\beta^2}$
7. $e^{-at} \sin \beta t$	$\frac{\beta}{(S+a)^2+\beta^2}$
8. $e^{-at} \cos \beta t$	$\frac{(S+a)}{(S+a)^2+\beta^2}$
9. $\sin^2(at)$	$\frac{2a^2(s^2+4a^2)}{S}$, $Re\ S > 2\ \lvert \mathrm{lm} \rvert$

$f(t)$	$F(s)$
10. $\cos^2(at)$	$\frac{s^2 + 2a^2}{S(S^2 + 4a^2)}$, $Re\ S > 2\ \lvert \text{lm} \rvert$
11. $\sin \sqrt{at}$	$\frac{(\pi a/2) \exp(-a^2/4s)}{S^{3/2}}$, $Re\ S > 0$
12. $\cos \sqrt{at}$	$\frac{1}{s} + \frac{ia}{2} \sqrt{\pi} \frac{\exp(-a^2/4S)}{S^{3/2}} \cdot Erf\left(\frac{ia}{2\sqrt{s}}\right.$, $Re\ S > 0$
13. $\frac{\sin at}{t}$	$\tan^{-1} \frac{a}{s}$, $Re\ S > \lvert \text{lm} \rvert$
14. $\frac{1 - \cos at}{t}$	$\frac{1}{2} \ln \left(1 + \frac{a^2}{s^2}\right)$, $Re\ S > \lvert \text{lm} \rvert$
15. $\frac{\sin at}{\sqrt{t}}$	$\sqrt{\frac{\pi}{2}} (S^2 + a^2)^{-1/2}[(s^2 + a^2)^{1/2} - S]^{1/2}$, $Re\ S > \lvert \text{lm} \rvert$
16. $\frac{\cos at}{\sqrt{t}}$	$\sqrt{\frac{\pi}{2}} (S^2 + a^2)^{-1/2}[(s^2 + a^2)^{1/2} + S]^{1/2}$, $Re\ S > \lvert \text{lm} \rvert$
17. $\frac{\sin^2 at}{t}$	$\frac{1}{4} \ln \left(1 + \frac{4a^2}{S^2}\right)$, $Re\ S > 2\ \lvert \text{lm} \rvert$
18. te^{-at}	$\frac{1}{(S + a)^2}$, $Re(S + a) > 0$
19. $f(t)e^{-at}$	$F(S + a)$
20. $t^{-1}(e^{-at} - e^{-bt})$	$\ln \frac{S + b}{S + a}$, $Re\ S > \max(-Re\ a, -Re\ b)$
21. $(1 - e^{-t/a})^n$, $n = 0, 1, 2, \ldots$	$\frac{n!}{S^{(1+aS)}n}$, $Re\ S > 0$, $Re \left(S + \frac{n}{a}\right) > 0$
22. $\frac{e^{-a/t}}{\sqrt{t}}$, $Re\ a \geq 0$	$\sqrt{\frac{\pi}{S}} \exp(-2\sqrt{as})$, $Re\ S > 0$
23. $\ln t$	$-\frac{\gamma + \ln S}{S}$, $Re\ S > 0$
24. $(\ln t)^2$	$\frac{\pi^2/\sigma + (\gamma + \ln S)^2}{S}$, $Re\ S > 0$
25. $t(1 - \ln t)$	$\frac{\gamma + \ln S}{S^2}$, $Re\ S > 0$
26. $\frac{\ln t}{t}$	$-\sqrt{\frac{\pi}{S}} [\gamma + \ln 4S]$, $Re\ S > 0$
27. $\sinh at$	$\frac{a}{S^2 - a^2}$, $Re\ S > \lvert Re\ a \rvert$
28. $\cosh at$	$\frac{S}{S^2 - a^2}$, $Re\ S > \lvert Re\ a \rvert$

	$f(t)$	$F(s)$
29.	$\dfrac{\sinh at}{t}$	$\dfrac{1}{2}\ln\dfrac{S+a}{S-a}$, $Re\ S > \lvert Re\ a\rvert$
30.	$\dfrac{1-\cosh at}{t}$	$\dfrac{1}{2}\ln\left(1-\dfrac{a^2}{S^2}\right)$, $Re\ S > \lvert Re\ a\rvert$
31.	$\dfrac{\sinh^2 at}{t}$	$-\dfrac{1}{2}\ln\left(1-\dfrac{4a^2}{S^2}\right)$, $Re\ S > 2\lvert Re\ a\rvert$
32.	$\cosh^2 at$	$\dfrac{S^2-2a^2}{S(S^2-4a^2)}$
33.	$\dfrac{\sinh^2 \sqrt{at}}{t}$	$\dfrac{1}{2}\sqrt{\dfrac{\pi}{S}}\left(\dfrac{a^2}{e^{S-1}}\right)$, $Re\ S > 0$
34.	$\dfrac{\cosh^2 st}{t}$	$\dfrac{1}{2}\sqrt{\dfrac{\pi}{S}}\left(\dfrac{a}{e^{S+1}}\right)$, $Re\ S > 0$
35.	$1-\dfrac{t+T}{T}e^{-t/T}$	$\dfrac{1}{S(1+TS)^2}$
36.	$\left[\dfrac{TW_n}{1+T^2W_n^2}e^{-t/T}+\dfrac{1}{1+T^2W_n^2}\sin(W_nt-Q)\right]$, where $Q=\tan^{-1}W_nT$	$\dfrac{W_n}{(1+TS)(S^2+W_n^2)}$
37.	$\left[t-\dfrac{2\gamma}{W_n}+\dfrac{1}{W_n\sqrt{1-\gamma^2}}e^{-W_nt}\sin(W_n\sqrt{1-\gamma^2}\,t-Q)\right]$, where $Q=2\tan^{-1}\dfrac{\sqrt{1-\gamma^2}}{-\gamma}$	$\dfrac{W_n^2}{S^2(S^2+2\gamma W_nS+W_n^2)}$
38.	$\left[1+\dfrac{TS_n^2(a-T)}{1+T^2W_n^2}e^{-t/T}-\sqrt{\dfrac{1+a^2W_n^2}{1+T^2W_n^2}}\cos(W_nt+Q)\right]$, where $Q=\tan^{-1}aW_n-\tan^{-1}W_nT$	$\dfrac{W_n^2(1+as)}{S(1+TS)(S^2+W_n^2)}$
39.	$J_0(at)$	$\dfrac{1}{\sqrt{S^2-a^2}}$
40.	$\ln tJ_0(at)$	$\dfrac{-\gamma-\ln 2+\ln[S+(S^2+a^2)^{1/2}]-\ln(S^2+a^2)}{S^2+a^2}$, $Re\ S > 0$
41.	$J_0(at)J_0(bt)$	$\dfrac{2}{\pi}[S^2+(a+b)^2]^{-1/2}\cdot K[2\sqrt{ab}\,[S^2+(a+b)^2]^{-1/2}$, $Re\ S > 0$
42.	$f\left(\dfrac{t}{a}\right)$	$aF(as)$
43.	$\dfrac{df(t)}{dt}$	$SF(S)-f(0^-)$

$f(t)$	$F(s)$
44. $\int_0^t f_1(t-\tau)f_2(\tau)\,d\tau$	$F_1(S)F_2(S)$
45. $f(t-a)U_{-1}(t-a)$	$e^{-as}F(S)$
46. $\int f(t)\,dt$	$\dfrac{F(S)}{S}+\dfrac{f^{-1}(0^+)}{S}$
47. $\dfrac{d^n f(t)}{dt^n} = S^n \cdot F(S) - \sum_{f=1}^{n} S^{(l-1)} \cdot f^{(n-1)}(0^-)$, where $f^{(n-1)}(0^-)$ is the initial value of the $(n-1)$th derivative of $f(t)$ at $t=0^-$ If any $f(t)$ is continuous at $t=0$, then $f(0^+)=f(0^-)$. However, if discontinuity exists at $t=0$, then $f(0^-)$ must be used.	
48. $\dfrac{\partial}{\partial a} f(t, a)$	$\dfrac{\partial}{\partial a} F(S, a)$
49. $t^n f(t)$	$(-1)^n \dfrac{d^n}{ds^n} F(S)$
50. $\int_{a_0}^{a} f(t, a)\,da$	$\int_{a_0}^{a} F(S, a)\,da$, $Re\, f(t) = Re\, F(S)$ $Im\, f(t) = Im\, F(S)$
51. $f_1(t)f_2(t) = \sum_{k=1}^{q} \dfrac{A_1(S_k)}{B_1'(S_k)} \cdot F_2(S-S_k)$, provided $F_1(S)$ is rational function, where $F_1(S) = A_1(S)/B_1(S)$ and has only q first-order poles. S_k are the poles of $F_1(S)$.	
52. $f_1(t)f_2(t) = \sum_{k=1}^{n}\sum_{j=1}^{M_k} \dfrac{(-1)^{M_k-j}\cdot K_{kj}}{(M_k-j)!}\left[\dfrac{d^{M_k-j}}{dS^{M_k-j}} F_2(S)\right]_{S=S-S_k}$, where $K_{kj} = \dfrac{1}{(j-1)!}\left[\dfrac{d^{j-1}}{dS^{j-1}}(S-S_k)^{M_k}\cdot F_1(S)\right]_{S=S_k}$ provided $F_1(S)$ is a rational function having n poles, where the pole S_k has an order of M_k.	
53. $\lim_{t\to\infty} f(t) = \lim_{s\to 0} SF(S)$	
54. $\lim_{t\to 0} f(t) = \lim_{s\to\infty} SF(S)$	
55. $\lim_{a\to a_0} f(t, a) = \lim_{a\to a_0} F(S, a)$	
56. $\dfrac{f(t)}{t}$	$\int_S^{\infty} F(S)\,ds$
57. $\int_{a_0}^{a} f(t, a)\,da$	$\int_{a_0}^{a} F(S, a)\,da$
58. $Re\, f(t)$ $Im\, F(t)$	$Re\, F(S)$ $Im\, F(S)$

Appendix C [2, 3, 7]

Some Fourier Transforms and Properties

C.1. FOURIER TRANSFORM OF SELECTED FUNCTIONS[1]

$f(t)$	$F(\omega)$
1. $U_0(t)$	1
2. $U_0(t - t_0)$	$e^{-j\omega t_0}$
3. $U_0'(t)$	$j\omega$
4. $U_{-1}(t)$	$U_0(\omega) + \dfrac{1}{j\omega}$
5. $U_{-1}(t - t_0)$	$U_0(\omega) + \dfrac{1}{j\omega} e^{-j\omega t_0}$
6. 1	$2\pi U_0(\omega)$
7. t	$2\pi j U_0'(\omega)$
8. t^n	$2\pi j^n U_0^{(n)}(\omega)$
9. $e^{j\omega_0 t}$	$2\pi\ U_0(\omega - \omega_0)$
10. $\cos \omega_0 t$	$\pi[U_0(\omega - \omega_0) + U_0(\omega + \omega_0)]$
11. $\sin \omega_0 t$	$-j\pi[U_0(\omega - \omega_0) - U_0(\omega + \omega_0)]$
12. $\sin \omega_0 t\ U_{-1}(t)$	$\dfrac{\omega_0}{\omega_0^2 - \omega^2} + \dfrac{\pi}{2j}[U_0(\omega - \omega_0) - U_0(\omega + \omega_0)]$
13. $\omega_0 t\ U_{-1}(t)$	$\dfrac{j\omega}{\omega_0^2 - \omega^2} + \dfrac{\pi}{2}[U_0(\omega - \omega_0) + U_0(\omega + \omega_0)]$

[1] OUTLINE OF FOURIER ANALYSIS, by H. P. HSU, 1967, pp. 339 and 340. © 1967 Simon & Schuster, Inc., New York.

$f(t)$	$F(\omega)$
14. $t\, U_{-1}(t)$	$j\pi U_0'(\omega) - \dfrac{1}{\omega^2}$
15. t^{-1}	$\pi j - 2\pi j\, U_{-1}(\omega)$
16. t^{-n}	$\dfrac{(-jW)^{n-1}}{(n-1)!}[\pi j - 2\pi j\, U_{-1}(\omega)]$
17. $U_0 T = \sum_{n=-\infty} U_0(t - nT)$,	$\sum_{n=-\infty} U_0(\omega - n_0^{\omega})$

where

$$\omega_0 = 2\pi/T$$
$$U_0 = \text{a unit impulse}$$
$$U_{-1}(t) = \text{a step function}$$

C.2. FORMULAS AND PROPERTIES OF FOURIER TRANSFORM[2]

1. Transform pair:

$$G(\omega) = {}^{\infty}\!\int_{-\infty} f(t)\, e^{-j\omega t}\, dt$$

$$f(t) = \frac{1}{2\pi}\, {}^{\infty}\!\int_{-\infty} (\omega) e^{j\omega t}\, d\omega$$

or

$$f(t) = \frac{1}{\pi}\, {}^{\infty}\!\int_{0} S(\omega) \cos[\omega t + Q(\omega)]\, d\omega$$

where

$$S(\omega) = \sqrt{a^2(\omega) + b^2(\omega)}$$

$$Q(\omega) = -\tan^{-1}\frac{b(\omega)}{a(\omega)}$$

$$a(\omega) = \frac{1}{\pi}\, {}^{\infty}\!\int_{-\infty} f(t) \cos \omega t\, dt$$

$$b(\omega) = \frac{1}{\pi}\, {}^{\infty}\!\int_{-\infty} f(t) \sin \omega t\, dt$$

a. If $G(\omega)$ is the transform of $f(t)$, then $G(-\omega)$ is the transform of $f(-t)$.
b. If $G(\omega)$ is the transform of $f(t)$, then $G(\pm\omega)$ is the transform of $\bar{f}(\pm t)$.

[2] Reprinted, with permission from *The Mathematics of Circuit Analysis* by E. A. Guillemin, 5th. Edition, 1958, pp. 524–527, John Wiley & Sons, Inc.

c. If $G(\omega)$ is the transform of $f(t)$, then
$f(\pm\omega)$ is the transform of $2\pi G(\pm t)$.

d. If $G(\omega)$ is the transform of $f(t)$, then
$(1/a)G(\omega/a)$ is the transform of $f(at)$.

e. If $G(\omega)$ is the transform of $f(t)$, then
$G(\omega \pm \omega_0)$ is the transform of $f(t) \cdot e^{\pm j\omega_0 t}$.

f. If $G(\omega)$ is the transform of $f(t)$, then
$G(\omega) \cdot e^{\pm j\omega t_0}$ is the transform of $f(t \pm t_0)$.

g. If $G(\omega)$ is the transform of $f(t)$, then
$G(j\omega)^n/G(\omega)$ is the transform of the nth derivative of $f(t)$.

h. If $G(\omega)$ is the transform of $f(t)$, then

$G(\omega)/j\omega$ is the transform of $\int_{-\infty}^{t} f(\theta)\, d\theta$.

i. If $G(\omega)$ is the transform of $f(t)$, then

$\int_{-\infty}^{\omega} G(\nu)\, d\nu$ is the transform of $f(t)/-jt$.

Appendix D

Conversion Factors*

To Convert	Multiply By	To Obtain
	A	
abamperes	$1. \times 10^{1}$	amperes
abcoulombs	2.998×10^{10}	statcoulombs
abfarads	$1. \times 10^{9}$	farads
abfarads	$1. \times 10^{15}$	microfarads
abhenries	$1. \times 10^{-9}$	henries
abhenries	$1. \times 10^{-6}$	millihenries
abohms	$1. \times 10^{-9}$	ohms
abohms	$1. \times 10^{-15}$	megohms
abvolts	$1. \times 10^{-8}$	volts
acres	$1. \times 10^{1}$	sq. chains (gunters)
acres	1.60×10^{2}	rods
acres	$1. \times 10^{5}$	sq. links
acres	4.047×10^{-1}	hectares or sq. hectometers
acres	4.35×10^{4}	sq. ft.
acres	4.047×10^{3}	sq. meters
acres	1.562×10^{-3}	sq. miles
acres	4.840×10^{3}	sq. yards
acre-feet	4.356×10^{4}	cu. feet
acre-feet	3.259×10^{5}	gallons
amperes/sq. cm.	6.452	amps/sq. in.
amperes/sq. cm.	$1. \times 10^{4}$	amps/sq. meter
amperes/sq. in.	1.550×10^{-1}	amps/sq. cm.
amperes/sq. in.	1.550×10^{3}	amps/sq. meter
amperes/sq. meter	1.0×10^{-4}	amps/sq. cm.

To Convert	Multiply By	To Obtain
amperes/sq. meter	6.452×10^{-4}	amps/sq. in.
ampere-hours	3.600×10^{3}	coulombs
ampere-hours	3.731×10^{-2}	faradays
ampere-turns	1.257	gilberts
ampere-turns/cm.	2.540	amp-turns/in.
ampere-turns/cm.	$1. \times 10^{2}$	amp-turns/meter
ampere-turns/in.	3.937×10^{-1}	amp-turns/cm.
ampere-turns/in.	3.937×10^{1}	amp-turns/meter
ampere-turns/in.	4.950×10^{-1}	gilberts/cm.
ampere-turns/meter	$1. \times 10^{-2}$	amp-turns/cm.
ampere-turns/meter	2.54×10^{-2}	amp-turns/in.
ampere-turns/meter	1.257×10^{-2}	gilberts/cm.
angstrom unit	3.937×10^{-9}	inches
angstrom unit	$1. \times 10^{-10}$	meters
angstrom unit	$1. \times 10^{-4}$	microns or (mu)
ares	2.471×10^{-2}	acres (u.s.)
ares	1.196×10^{2}	sq. yards
ares	$1. \times 10^{2}$	sq. meters
astronomical unit	1.495×10^{8}	kilometers
atmospheres	7.348×10^{-3}	tons/sq. in.
atmospheres	1.058	tons/sq. foot
atmospheres	7.6×10^{1}	cms. of mercury (at 0°C.)
atmospheres	3.39×10^{1}	ft.of water (at 4°C.)
atmospheres	2.992×10^{1}	in. of mercury (at 0°C.)
atmospheres	7.6×10^{-1}	meters of mercury (at 0°C.)
atmospheres	7.6×10^{2}	millimeters of mercury (at 0°C.)
atmospheres	1.0333	kgs./sq. cm.
atmospheres	1.0333×10^{4}	kgs./sq. meter
atmospheres	1.47×10^{1}	pounds/sq. in.
	B	
barrels (u.s., dry)	3.281	bushels
barrels (u.s., dry)	7.056×10^{3}	cu. inches
barrels (u.s., dry)	1.05×10^{2}	quarts (dry)
barrels (u.s. liquid)	3.15×10^{1}	gallons
barrels (oil)	4.2×10^{1}	gallons (oil)
bars	9.869×10^{-1}	atmospheres
bars	$1. \times 10^{6}$	dynes/sq. cm.
bars	1.020×10^{4}	kgs./sq. meter
bars	2.089×10^{3}	pounds/sq. ft.
bars	1.45×10^{1}	pounds/sq. in.
barye	1.00	dynes/sq. cm.
bolt (u.s., cloth)	3.6576×10^{1}	meters
btu	1.0409×10^{1}	liter-atmospheres
btu	1.0550×10^{10}	ergs
btu	7.7816×10^{2}	foot-pounds
btu	2.52×10^{2}	gram-calories
btu	3.927×10^{-4}	horsepower-hours
btu	1.055×10^{3}	joules
btu	2.52×10^{-1}	kilogram-calories
btu	1.0758×10^{2}	kilogrammeters

To Convert	Multiply By	To Obtain
btu	2.928×10^{-4}	kilowatt-hours
btu/hr.	2.162×10^{-1}	ft.-pounds/sec.
btu/hr.	7.0×10^{-2}	gram-cal./sec.
btu/hr.	3.929×10^{-4}	horsepower
btu/hr.	2.931×10^{-1}	watts
btu/min.	1.296×10^{1}	ft.-pounds/sec.
btu/min.	2.356×10^{-2}	horsepower
btu/min.	1.757×10^{-2}	kilowatts
btu/min.	1.757×10^{1}	watts
btu/sq. ft./min.	1.22×10^{-1}	watts/sq. in.
bucket (br. dry)	1.8184×10^{4}	cubic cm.
bushels	1.2445	cubic ft.
bushels	2.1504×10^{3}	cubic in.
bushels	3.524×10^{-2}	cubic meters
bushels	3.524×10^{1}	liters
bushels	4.0	pecks
bushels	6.4×10^{1}	pints (dry)
bushels	3.2×10^{1}	quarts (dry)
	C	
calories, gram (mean)	3.9685×10^{-3}	btu (mean)
candle/sq. cm.	3.146	lamberts
candle/sq. in.	4.870×10^{-1}	lamberts
centares	1.0	sq. meters
centigrade (degrees)	$(°C \times 9/5) + 32$	fahrenheit (degrees)
centigrade (degrees)	$°C + 273.18$	kelvin (degrees)
centigrams	$1. \times 10^{-2}$	grams
centiliters	3.382×10^{-1}	ounce (fluid) u.s.
centiliters	6.103×10^{-1}	cubic in.
centiliters	2.705	drams
centiliters	1.0×10^{-2}	liters
centimeters	3.281×10^{-2}	feet
centimeters	3.937×10^{-1}	inches
centimeters	$1. \times 10^{-5}$	kilometers
centimeters	$1. \times 10^{-2}$	meters
centimeters	6.214×10^{-6}	miles
centimeters	$1. \times 10^{1}$	millimeters
centimeters	3.937×10^{2}	mils
centimeters	1.094×10^{-2}	yards
centimeters	$1. \times 10^{4}$	microns
centimeters	$1. \times 10^{8}$	angstrom units
centimeter-dynes	1.020×10^{-3}	cn-grams
centimeter-dynes	1.020×10^{-8}	meter-kgs.
centimeter-dynes	7.376×10^{-8}	pound-ft.
centimeter-grams	9.807×10^{2}	cm.-dynes
centimeter-grams	$1. \times 10^{-5}$	meter-kgs.
centimeter-grams	7.233×10^{-5}	pound-ft.
centimeters of mercury	1.316×10^{-2}	atmospheres
centimeters of mercury	4.461×10^{-1}	ft. of water
centimeters of mercury	1.36×10^{2}	kgs./sq. meter
centimeters of mercury	2.785×10^{1}	pounds/sq. ft.

To Convert	Multiply By	To Obtain
centimeters of mercury	1.934×10^{-1}	pounds/sq. in.
centimeters/sec.	1.969	feet/min.
centimeters/sec.	3.281×10^{-2}	feet/sec.
centimeters/sec.	3.6×10^{-2}	kilometers/hr.
centimeters/sec.	1.943×10^{-2}	knots
centimeters/sec.	6.0×10^{-1}	meters/min.
centimeters/sec.	2.237×10^{-2}	miles/hr.
centimeters/sec.	3.728×10^{-4}	miles/min.
centimeters/sec./sec.	3.281×10^{-2}	ft./sec./sec.
centimeters/sec./sec.	3.6×10^{-2}	kms./hr./sec.
centimeters/sec./sec.	1.0×10^{-2}	meters/sec./sec.
centimeters/sec./sec.	2.237×10^{-2}	miles/hr./sec.
centipoise	1.0×10^{-2}	gr./cm.-sec.
centipoise	6.72×10^{-4}	pound/ft.-sec.
centipoise	2.4	pound/ft.-hr.
chains (gunters)	7.92×10^{2}	inches
chains (gunters)	2.012×10^{1}	meters
chains (gunters)	2.2×10^{1}	yards
circular mils	5.067×10^{-6}	sq. cm.
circular mils	7.854×10^{-1}	sq. mils
circular mils	7.854×10^{-7}	sq. inches
circumference	6.283	radians
cords	8.0	cord ft.
cord ft.	1.6×10^{1}	cubic ft.
coulombs	2.998×10^{9}	statcoulombs
coulombs	1.036×10^{-5}	faradays
coulombs/sq. cm.	6.452	coulombs/sq. in.
coulombs/sq. cm.	1.0×10^{4}	coulombs/sq. meter
coulombs/sq. in.	1.550×10^{-1}	coulombs/sq. cm.
coulombs/sq. in.	1.550×10^{3}	coulombs/sq. meter
coulombs/sq. meter	1.0×10^{-4}	coulombs/sq. cm.
coulombs/sq. meter	6.452×10^{-4}	coulombs/sq. in.
cubic centimeters	3.531×10^{-5}	cubic ft.
cubic centimeters	6.102×10^{-2}	cubic in.
cubic centimeters	1.0×10^{-6}	cubic meters
cubic centimeters	1.308×10^{-6}	cubic yards
cubic centimeters	2.642×10^{-4}	gallons (u.s. liquid)
cubic centimeters	1.0×10^{-3}	liters
cubic centimeters	2.113×10^{-3}	pints (u.s. liquid)
cubic centimeters	1.057×10^{-3}	quarts (u.s. liquid)
cubic feet	8.036×10^{-1}	bushels (dry)
cubic feet	2.8320×10^{4}	cu. cms.
cubic feet	1.728×10^{3}	cu. inches
cubic feet	2.832×10^{-2}	cu. meters
cubic feet	3.704×10^{-2}	cu. yards
cubic feet	7.48052	gallons (u.s. liquid)
cubic feet	2.832×10^{1}	liters
cubic feet	5.984×10^{1}	pints (u.s. liquid)
cubic feet	2.992×10^{1}	quarts (u.s. liquid)
cubic feet/min.	4.72×10^{2}	cu. cms./sec.
cubic feet/min.	1.247×10^{-1}	gallons/sec.

To Convert	Multiply By	To Obtain
cubic feet/min.	4.720×10^{-1}	liters/sec.
cubic feet/min.	6.243×10^{1}	pounds water/min.
cubic feet/sec.	6.46317×10^{-1}	million gals./day
cubic feet/sec.	4.48831×10^{2}	gallons/min.
cubic inches	1.639×10^{1}	cu. cms.
cubic inches	5.787×10^{-4}	cu. ft.
cubic inches	1.639×10^{-5}	cu. meters
cubic inches	2.143×10^{-5}	cu. yards
cubic inches	4.329×10^{-3}	gallons
cubic inches	1.639×10^{-2}	liters
cubic inches	3.463×10^{-2}	pints (u.s. liquid)
cubic inches	1.732×10^{-2}	quarts (u.s. liquid)
cubic meters	2.838×10^{1}	bushels (dry)
cubic meters	1.0×10^{6}	cu. cms.
cubic meters	3.531×10^{1}	cu. ft.
cubic meters	6.1023×10^{4}	cu. inches
cubic meters	1.308	cu. yards
cubic meters	2.642×10^{2}	gallons (u.s. liquid)
cubic meters	1.0×10^{3}	liters
cubic meters	2.113×10^{3}	pints (u.s. liquid)
cubic meters	1.057×10^{3}	quarts (u.s. liquid)
cubic yards	7.646×10^{5}	cu. cms.
cubic yards	2.7×10^{1}	cu. ft.
cubic yards	4.6656×10^{4}	cu. inches
cubic yards	7.646×10^{-1}	cu. meters
cubic yards	2.02×10^{2}	gallons (u.s. liquid)
cubic yards	7.646×10^{2}	liters
cubic yards	1.6159×10^{3}	pints (u.s. liquid)
cubic yards	8.079×10^{2}	quarts (u.s. liquid)
cubic yards/min.	4.5×10^{-1}	cubic ft./sec.
cubic yards/min.	3.367	gallons/sec.
cubic yards/min.	1.274×10^{1}	liters/sec.
	D	
daltons	1.650×10^{-24}	grams
days	8.64×10^{4}	seconds
days	1.44×10^{3}	minutes
days	2.4×10^{1}	hours
decigrams	1.0×10^{-1}	grams
deciliters	1.0×10^{-1}	liters
decimeters	1.0×10^{-1}	meters
degrees (angle)	1.111×10^{-2}	quadrants
degrees (angle)	1.745×10^{-2}	radians
degrees (angle)	3.6×10^{3}	seconds
degrees/sec.	1.745×10^{-2}	radians/sec.
degrees/sec.	1.667×10^{-1}	revolutions/min.
degrees/sec.	2.778×10^{-3}	revolutions/sec.
dekagrams	1.0×10^{1}	grams
dekaliters	1.0×10^{1}	liters
dekameters	1.0×10^{1}	meters
drams (apoth. or troy)	1.3714×10^{-1}	ounces (avdp.)

To Convert	Multiply By	To Obtain
drams (apoth. or troy)	1.25×10^{-1}	ounces (troy)
drams (u.s. fluid or apoth.)	3.6967	cubic cm.
drams	1.7718	grams
drams	2.7344×10^{1}	grains
drams	6.25×10^{-2}	ounces
dynes/sq. cm.	1.0×10^{-2}	ergs/sq. millimeter
dynes/sq. cm.	9.869×10^{-7}	atmospheres
dynes/sq. cm.	2.953×10^{-5}	in. of mercury (at 0°C.)
dynes/sq. cm.	4.015×10^{-4}	in. of water (at 4°C.)
dynes	1.020×10^{-3}	grams
dynes	1.0×10^{-7}	joules/cm.
dynes	1.0×10^{-5}	joules/meter (newtons)
dynes	1.020×10^{-6}	kilograms
dynes	7.233×10^{-5}	poundals
dynes	2.248×10^{-6}	pounds
dynes/sq. cm.	1.0×10^{-6}	bars
	E	
ell	1.1430×10^{2}	cm.
ell	4.5×10^{1}	inches
em, pica	1.67×10^{-1}	inch
em, pica	4.233×10^{-1}	cm.
erg/sec.	1.0	dyne-cm./sec.
ergs	9.486×10^{-11}	btu
ergs	1.0	dyne-centimeters
ergs	7.376×10^{-8}	foot-pounds
ergs	2.389×10^{-8}	gram-calories
ergs	1.020×10^{-3}	gram-cms.
ergs	3.7250×10^{-14}	horsepower-hrs.
ergs	1.0×10^{-7}	joules
ergs	2.389×10^{-11}	kg.-calories
ergs	1.020×10^{-8}	kg.-meters
ergs	2.773×10^{-14}	kilowatt-hrs.
ergs	2.773×10^{-11}	watt-hrs.
ergs/sec.	5.668×10^{-9}	btu/min.
ergs/sec.	4.426×10^{-6}	ft.-lbs./min.
ergs/sec.	7.3756×10^{-8}	ft.-lbs./sec.
ergs/sec.	1.341×10^{-10}	horsepower
ergs/sec.	1.433×10^{-9}	kg-calories/min.
ergs/sec.	$1. \times 10^{-10}$	kilowatts
	F	
farads	$1. \times 10^{6}$	microfarads
faraday/sec.	9.65×10^{4}	ampere (absolute)
faradays	2.68×10^{1}	ampere-hours
faradays	9.649×10^{4}	coulombs
fathoms	1.8288	meters
fathoms	6.0	feet
feet	3.048×10^{1}	centimeters
feet	3.048×10^{-4}	kilometers

To Convert	Multiply By	To Obtain
feet	3.048×10^{-1}	meters
feet	1.645×10^{-4}	miles (naut.)
feet	1.894×10^{-4}	miles (stat.)
feet	3.048×10^{2}	millimeters
feet	1.2×10^{4}	mils
feet of water	2.95×10^{-2}	atmospheres
feet of water	8.826×10^{-1}	in. of mercury
feet of water	3.048×10^{-2}	kgs./sq. cm.
feet of water	3.048×10^{2}	kgs./sq. meter
feet of water	6.243×10^{1}	pounds/sq. ft.
feet of water	4.335×10^{-1}	pounds/sq. in.
feet/min.	5.080×10^{-1}	cms./sec.
feet/min.	1.667×10^{-2}	feet/sec.
feet/min.	1.829×10^{-2}	kms./hr.
feet/min.	3.048×10^{-1}	meters/min.
feet/min.	1.136×10^{-2}	miles/hr.
feet/sec.	3.048×10^{1}	cms./sec.
feet/sec.	1.097	kms./hr.
feet/sec.	5.921×10^{-1}	knots
feet/sec.	1.829×10^{1}	meters/min.
feet/sec.	6.818×10^{-1}	miles/hr.
feet/sec.	1.136×10^{-2}	miles/min.
feet/sec./sec.	3.048×10^{1}	cms./sec./sec.
feet/sec./sec.	1.097	kms./hr./sec.
feet/sec./sec.	3.048×10^{-1}	meters/sec./sec.
feet/sec./sec.	6.818×10^{-1}	miles/hr./sec.
feet/100 feet	1.0	per cent grade
foot-candle	1.0764×10^{1}	lumen/sq. meter
foot-candle	1.0764×10^{1}	lux
foot-pounds	1.286×10^{-3}	btu
foot-pounds	1.356×10^{7}	ergs
foot-pounds	3.241×10^{-1}	gram-calories
foot-pounds	5.050×10^{-7}	horsepower-hrs.
foot-pounds	1.356	joules
foot-pounds	3.241×10^{-4}	kg.-calories
foot-pounds	1.383×10^{-1}	kg.-meters
foot-pounds	3.766×10^{-7}	kilowatt-hrs.
foot-pounds/min.	1.286×10^{-3}	btu/min.
foot-pounds/min.	1.667×10^{-2}	foot-pounds/sec.
foot-pounds/min.	3.030×10^{-5}	horsepower
foot-pounds/min.	3.241×10^{-4}	kg.-calories/min.
foot-pounds/min.	2.260×10^{-5}	kilowatts
foot-pounds/sec.	4.6263	btu/hr.
foot-pounds/sec.	7.717×10^{-2}	btu/min.
foot-pounds/sec.	1.818×10^{-3}	horsepower
foot-pounds/sec.	1.945×10^{-2}	kg.-calories/min.
foot-pounds/sec.	1.356×10^{-3}	kilowatts
furlongs	1.25×10^{-1}	miles (u.s.)
furlongs	4.0×10^{1}	rods
furlongs	6.6×10^{2}	feet
furlongs	2.0117×10^{2}	meters

To Convert	Multiply By	To Obtain
	G	
gallons	3.785×10^{3}	cu. cms.
gallons	1.337×10^{-1}	cu. feet
gallons	2.31×10^{2}	cu. inches
gallons	3.785×10^{-3}	cu. meters
gallons	4.951×10^{-3}	cu. yards
gallons	3.785	liters
gallons (liq. br. imp.)	1.20095	gallons (u.s. liquid)
gallons (u.s.)	8.3267×10^{-1}	gallons (imp.)
gallons of water	8.337	pounds of water
gallons/min.	2.228×10^{-3}	cu. feet/sec.
gallons/min.	6.308×10^{-2}	liters/sec.
gallons/min.	8.0208	cu. feet/hr.
gausses	6.452	lines/sq. in.
gausses	1.0×10^{-8}	webers/sq. cm.
gausses	6.452×10^{-8}	webers/sq. in.
gausses	1.0×10^{-4}	webers/sq. meter
gausses	7.958×10^{-1}	amp.-turn/cm.
gausses	1.0	gilbert/cm.
gilberts	7.958×10^{-1}	ampere-turns
gilberts/cm.	7.958×10^{-1}	ampere-turns/cm.
gilberts/cm.	2.021	ampere-turns/in.
gilberts/cm.	7.958×10^{1}	ampere-turns/meter
gills (british)	1.4207×10^{2}	cubic cm.
gills (u.s.)	1.18295×10^{2}	cubic cm.
gills (u.s.)	1.183×10^{-1}	liters
gills (u.s.)	2.5×10^{-1}	pints (liq.)
grade	1.571×10^{-2}	radian
grains	3.657×10^{-2}	drams (avdp.)
grains (troy)	1.0	grains (avdp.)
grains (troy)	6.48×10^{-2}	grams
grains (troy)	2.0833×10^{-3}	ounces (avdp.)
grains (troy)	4.167×10^{-2}	pennyweight (troy)
grains/u.s. gallon	1.7118×10^{1}	parts/million
grains/u.s. gallon	1.4286×10^{2}	pounds/million gallons
grains/imp. gallon	1.4286×10^{1}	parts/million
grams	9.807×10^{2}	dynes
grams	1.543×10^{1}	grains (troy)
grams	9.807×10^{-5}	joules/cm.
grams	9.807×10^{-3}	joules/meter (newtons)
grams	1.0×10^{-3}	kilograms
grams	1.0×10^{3}	milligrams
grams	3.527×10^{-2}	ounces (avdp.)
grams	3.215×10^{-2}	ounces (troy)
grams	7.093×10^{-2}	poundals
grams	2.205×10^{-3}	pounds
grams/cm.	5.6×10^{-3}	pounds/in.
grams/cu. cm.	6.243×10^{1}	pounds/cu. ft.
grams/cu. cm.	3.613×10^{-2}	pounds/cu. in.
grams/cu. cm.	3.405×10^{-7}	pounds/mil-foot

To Convert	Multiply By	To Obtain
grams/liter	5.8417×10^{1}	grains/gal.
grams/liter	8.345	pounds/1,000 gal.
grams/liter	6.2427×10^{-2}	pounds/cu. ft.
grams/sq. cm.	2.0481	pounds/sq. ft.
gram-calories	3.9683×10^{-3}	btu
gram-calories	4.184×10^{7}	ergs
gram-calories	3.086	foot-pounds
gram-calories	1.5596×10^{-6}	horsepower-hrs.
gram-calories	1.162×10^{-6}	kilowatt-hrs.
gram-calories	1.162×10^{-3}	watt-hrs.
gram-calories/sec.	1.4286×10^{1}	btu/hr.
gram-centimeters	9.297×10^{-8}	btu
gram-centimeters	9.807×10^{2}	ergs
gram-centimeters	9.807×10^{-5}	joules
gram-centimeters	2.343×10^{-8}	kg.-calories
gram-centimeters	1.0×10^{-5}	kg.-meters
	H	
hand	1.016×10^{1}	cm.
hectares	2.471	acres
hectares	1.076×10^{5}	sq. feet
hectograms	1.0×10^{2}	grams
hectoliters	1.0×10^{2}	liters
hectometers	1.0×10^{2}	meters
hectowatts	1.0×10^{2}	watts
henries	1.0×10^{3}	millihenries
hogsheads (british)	1.0114×10^{1}	cubic ft.
hogsheads (u.s.)	8.42184	cubic ft.
hogsheads (u.s.)	6.3×10^{1}	gallons (u.s.)
horsepower	4.244×10^{1}	btu/min.
horsepower	3.3×10^{4}	foot-lbs./min.
horsepower	5.50×10^{2}	foot-lbs./sec.
horsepower (metric)	9.863×10^{-1}	horsepower
horsepower	1.014	horsepower (metric)
horsepower	1.068×10^{1}	kg.-calories/min.
horsepower	7.457×10^{-1}	kilowatts
horsepower	7.457×10^{2}	watts
horsepower (boiler)	3.352×10^{4}	btu/hr.
horsepower (boiler)	9.803	kilowatts
horsepower-hours	2.547×10^{3}	btu
horsepower-hours	2.6845×10^{13}	ergs
horsepower-hours	1.98×10^{6}	foot-lbs.
horsepower-hours	6.4119×10^{5}	gram-calories
horsepower-hours	2.684×10^{6}	joules
horsepower-hours	6.417×10^{2}	kg.-calories
horsepower-hours	2.737×10^{5}	kg.-meters
horsepower-hours	7.457×10^{-1}	kilowatt-hrs.
hours	4.167×10^{-2}	days
hours	5.952×10^{-3}	weeks
hours	3.6×10^{3}	seconds
hundredwgts(long)	1.12×10^{2}	pounds

To Convert	Multiply By	To Obtain
hundredwgts(long)	5.0×10^{-2}	tons (long)
hundredwgts(long)	5.08023×10^{1}	kilograms
hundredwgts(short)	4.53592×10^{-2}	tons (metric)
hundredwgts(short)	4.46429×10^{-2}	tons (long)
hundredwgts(short)	4.53592×10^{1}	kilograms
	I	
inches	2.540	centimeters
inches	2.540×10^{-2}	meters
inches	1.578×10^{-5}	miles
inches	2.54×10^{1}	millimeters
inches	1.0×10^{3}	mils
inches	2.778×10^{-2}	yards
inches	2.54×10^{8}	angstrom units
inches	5.0505×10^{-3}	rods
inches of mercury	3.342×10^{-2}	atmospheres
inches of mercury	1.133	feet of water
inches of mercury	3.453×10^{-2}	kgs./sq. cm.
inches of mercury	3.453×10^{2}	kgs./sq. meter
inches of mercury	7.073×10^{1}	pounds/sq. ft.
inches of mercury	4.912×10^{-1}	pounds/sq. in.
in. of water (at 4°C.)	2.458×10^{-3}	atmospheres
in. of water (at 4°C.)	7.355×10^{-2}	inches of mercury
in. of water (at 4°C.)	2.54×10^{-3}	kgs./sq. cm.
in. of water (at 4°C.)	5.781×10^{-1}	ounces/sq. in.
in. of water (at 4°C.)	5.204	pounds/sq. ft.
in. of water (at 4°C.)	3.613×10^{-2}	pounds/sq. in.
internat'l ampere	9.998×10^{-1}	absolute amp. (u.s.)
internat'l volt	1.00033	absolute volt (u.s.)
internat'l coulomb	9.99835×10^{-1}	absolute coulomb
	J	
joules	9.486×10^{-4}	btu
joules	1.0×10^{7}	ergs
joules	7.736×10^{-1}	foot-pounds
joules	2.389×10^{-4}	kg.-calories
joules	1.020×10^{-1}	kg.-meters
joules	2.778×10^{-4}	watt-hrs.
joules/cm.	1.020×10^{4}	grams
joules/cm.	1.0×10^{7}	dynes
joules/cm.	1.0×10^{2}	joules/meter (newtons)
joules/cm.	7.233×10^{2}	poundals
joules/cm.	2.248×10^{1}	pounds
	K	
kilograms	9.80665×10^{5}	dynes
kilograms	1.0×10^{3}	grams
kilograms	9.807×10^{-2}	joules/cm.
kilograms	9.807	joules/meter (newtons)
kilograms	7.093×10^{1}	poundals
kilograms	2.2046	pounds

To Convert	Multiply By	To Obtain
kilograms	9.842×10^{-4}	tons (long)
kilograms	1.102×10^{-3}	tons (short)
kilograms	3.5274×10^{1}	ounces (avdp.)
kilograms/cu. meter	1.0×10^{-3}	grams/cu. cm.
kilograms/cu. meter	6.243×10^{-2}	pounds/cu. ft.
kilograms/cu. meter	3.613×10^{-5}	pounds/cu. in.
kilograms/cu. meter	3.405×10^{-10}	pounds/mil-foot
kilograms/meter	6.72×10^{-1}	pounds/ft.
kilograms/sq. cm.	9.80665×10^{5}	dynes/sq. cm.
kilograms/sq. cm.	9.678×10^{-1}	atmospheres
kilograms/sq. cm.	3.281×10^{1}	feet of water
kilograms/sq. cm.	2.896×10^{1}	inches of mercury
kilograms/sq. cm.	2.048×10^{3}	pounds/sq. ft.
kilograms/sq. cm.	1.422×10^{1}	pounds/sq. in.
kilograms/sq. meter	9.678×10^{-5}	atmospheres
kilograms/sq. meter	9.807×10^{-5}	bars
kilograms/sq. meter	3.281×10^{-3}	feet of water
kilograms/sq. meter	2.896×10^{-3}	inches of mercury
kilograms/sq. meter	2.048×10^{-1}	pounds/sq. ft.
kilograms/sq. meter	1.422×10^{-3}	pounds/sq. in.
kilograms/sq. meter	9.80665×10^{1}	dynes/sq. cm.
kilograms/sq. mm.	1.0×10^{6}	kgs./sq. meter
kilogram-calories	3.968	btu
kilogram-calories	3.086×10^{3}	foot-pounds
kilogram-calories	1.558×10^{-3}	horsepower-hrs.
kilogram-calories	4.183×10^{3}	joules
kilogram-calories	4.269×10^{2}	kg.-meters
kilogram-calories	4.186	kilojoules
kilogram-calories	1.163×10^{-3}	kilowatt-hrs.
kilogram-calories/min.	5.143×10^{1}	ft.-lbs./sec.
kilogram-calories/min.	9.351×10^{-2}	horsepower
kilogram-calories/min.	6.972×10^{-2}	kilowatts
kilogram-meters	9.296×10^{-3}	btu
kilogram-meters	9.807×10^{7}	ergs
kilogram-meters	7.233	foot-pounds
kilogram-meters	9.807	joules
kilogram-meters	2.342×10^{-3}	kg.-calories
kilogram-meters	2.723×10^{-6}	kilowatt-hrs.
kilolines	1.0×10^{3}	maxwells
kiloliters	1.0×10^{3}	liters
kiloliters	1.308	cubic yards
kiloliters	3.5316×10^{1}	cubic feet
kiloliters	2.6418×10^{2}	gallons (u.s. liquid)
kilometers	1.0×10^{5}	centimeters
kilometers	3.281×10^{3}	feet
kilometers	3.937×10^{4}	inches
kilometers	1.0×10^{3}	meters
kilometers	6.214×10^{-1}	miles (statute)
kilometers	5.396×10^{-1}	miles (nautical)
kilometers	1.0×10^{6}	millimeters
kilometers	1.0936×10^{3}	yards

To Convert	Multiply By	To Obtain
kilometers/hr.	2.778×10^{1}	cms./sec.
kilometers/hr.	5.468×10^{1}	feet/min.
kilometers/hr.	9.113×10^{-1}	feet/sec.
kilometers/hr.	5.396×10^{-1}	knots
kilometers/hr.	1.667×10^{1}	meters/min.
kilometers/hr.	6.214×10^{-1}	miles/hr.
kilometers/hr./sec.	2.778×10^{1}	cms./sec./sec.
kilometers/hr./sec.	9.113×10^{-1}	ft./sec./sec.
kilometers/hr./sec.	2.778×10^{-1}	meters/sec./sec.
kilometers/hr./sec.	6.214×10^{-1}	miles/hr./sec.
kilowatts	5.692×10^{1}	btu/min.
kilowatts	4.426×10^{4}	foot-lbs./min.
kilowatts	7.376×10^{2}	foot-lbs./sec.
kilowatts	1.341	horsepower
kilowatts	1.434×10^{1}	kg.-calories/min.
kilowatts	1.0×10^{3}	watts
kilowatt-hrs.	3.413×10^{3}	btu
kilowatt-hrs.	3.6×10^{13}	ergs
kilowatt-hrs.	2.655×10^{6}	foot-lbs.
kilowatt-hrs.	8.5985×10^{5}	gram calories
kilowatt-hrs.	1.341	horsepower-hours
kilowatt-hrs.	3.6×10^{6}	joules
kilowatt-hrs.	8.605×10^{2}	kg.-calories
kilowatt-hrs.	3.671×10^{5}	kg.-meters
kilowatt-hrs.	3.53	pounds of water evaporated from and at 212°F.
kilowatt-hrs.	2.275×10^{1}	pounds of water raised from 62° to 212°F.
knots	6.076×10^{3}	feet/hr.
knots	1.852	kilometers/hr.
knots	1.0	nautical miles/hr.
knots	1.151	statute miles/hr.
knots	2.027×10^{3}	yards/hr.
knots	1.688	feet/sec.
knots	5.144×10^{1}	cm./sec.
	L	
lambert	3.183×10^{-1}	candle/sq. cm.
lambert	2.054	candle/sq. in.
league	3.0	miles (approx.)
light year	5.9×10^{12}	miles
light year	9.46091×10^{12}	kilometers
lines/sq. cm.	1.0	gausses
lines/sq. in.	1.55×10^{-1}	gausses
lines/sq. in.	1.55×10^{-9}	webers/sq. cm.
lines/sq. in.	1.0×10^{-8}	webers/sq. in.
lines/sq. in.	1.55×10^{-5}	webers/sq. meter
links (engineers)	1.2×10^{1}	inches
links (surveyors)	7.92	inches
liters	2.838×10^{-2}	bushels (u.s. dry)
liters	1.0×10^{3}	cu. cm.

To Convert	Multiply By	To Obtain
liters	3.531×10^{-2}	cu. ft.
liters	6.102×10^{1}	cu. inches
liters	1.0×10^{-3}	cu. meters
liters	1.308×10^{-3}	cu. yards
liters	2.642×10^{-1}	gallons (u.s. liquid)
liters	2.113	pints (u.s. liquid)
liters	1.057	quarts (u.s. liquid)
liters/min.	5.886×10^{-4}	cu. ft./sec.
liters/min.	4.403×10^{-3}	gals./sec.
$\log_{10} n$	2.303	ln n
ln n	4.343×10^{-1}	$\log_{10} n$
lumen	7.958×10^{-2}	spherical candle power
lumen/sq. ft.	1.0	foot-candles
lumen/sq. ft.	1.076×10^{1}	lumen-sq. meter
lux	9.29×10^{-2}	foot-candles
	M	
maxwells	1.0×10^{-3}	kilolines
maxwells	1.0×10^{-8}	webers
megalines	1.0×10^{4}	maxwells
megohms	1.0×10^{12}	microhms
megohms	1.0×10^{6}	ohms
megmhos/cubic cm.	1.0×10^{-3}	abmhos/cubic cm.
megmhos/cubic cm.	2.54	megmhos/cubic in.
megmhos/cubic cm.	1.662×10^{-1}	mhos/mil. ft.
megmhos/in. cube	3.937×10^{-1}	megmhos/cubic cm.
meters	1.0×10^{10}	angstrom units
meters	1.0×10^{2}	centimeters
meters	5.4681×10^{-1}	fathoms
meters	3.281	feet
meters	3.937×10^{1}	inches
meters	1.0×10^{-3}	kilometers
meters	5.400×10^{-4}	miles (nautical)
meters	6.214×10^{-4}	miles (statute)
meters	1.0×10^{3}	millimeters
meters	1.094	yards
meters/min.	1.667	cms./sec.
meters/min.	3.281	feet/min.
meters/min.	5.468×10^{-2}	feet/sec.
meters/min.	6.0×10^{-2}	kms./hr.
meters/min.	3.240×10^{-2}	knots
meters/min.	3.728×10^{-2}	miles/hr.
meters/sec.	1.968×10^{2}	feet/min.
meters/sec.	3.281	feet/sec.
meters/sec.	3.6	kilometers/hr.
meters/sec.	6.0×10^{-2}	kilometers/min.
meters/sec.	2.237	miles/hr.
meters/sec.	3.728×10^{-2}	miles/min.
meters/sec./sec.	1.0×10^{2}	cms./sec./sec.
meters/sec./sec.	3.281	ft./sec./sec.
meters/sec./sec.	3.6	kms./hr./sec.

To Convert	Multiply By	To Obtain
meters/sec./sec.	2.237	miles/hr./sec.
meter-kilograms	9.807×10^{7}	cm.-dynes
meter-kilograms	1.0×10^{5}	cm.-grams
meter-kilograms	7.233	pound-feet
microfarads	1.0×10^{-15}	abfarads
microfarads	1.0×10^{-6}	farads
microfarads	9.0×10^{5}	statfarads
micrograms	1.0×10^{-6}	grams
microhms	1.0×10^{3}	abohms
microhms	1.0×10^{-12}	megohms
microhms	1.0×10^{-6}	ohms
microliters	1.0×10^{-6}	liters
micromicrons	1.0×10^{-12}	meters
microns	1.0×10^{-6}	meters
miles (nautical)	6.076×10^{3}	feet
miles (nautical)	1.852	kilometers
miles (nautical)	1.852×10^{3}	meters
miles (nautical)	1.1516	miles (statute)
miles (nautical)	2.0254×10^{3}	yards
miles (statute)	1.609×10^{5}	centimeters
miles (statute)	5.280×10^{3}	feet
miles (statute)	6.336×10^{4}	inches
miles (statute)	1.609	kilometers
miles (statute)	1.609×10^{3}	meters
miles (statute)	8.684×10^{-1}	miles (nautical)
miles (statute)	1.760×10^{3}	yards
miles (statute)	1.69×10^{-13}	light years
miles/hr.	4.470×10^{1}	cms./sec.
miles/hr.	8.8×10^{1}	ft./min.
miles/hr.	1.467	ft./sec.
miles/hr.	1.6093	kms./hr.
miles/hr.	2.682×10^{-2}	kms./min.
miles/hr.	8.684×10^{-1}	knots
miles/hr.	2.682×10^{1}	meters/min.
miles/hr.	1.667×10^{-2}	miles/min.
miles/hr./sec.	4.47×10^{1}	cms./sec./sec.
miles/hr./sec.	1.467	ft./sec./sec.
miles/hr./sec.	1.6093	kms./hr./sec.
miles/hr./sec.	4.47×10^{-1}	meters/sec./sec.
miles/min.	2.682×10^{3}	cms./sec.
miles/min.	8.8×10^{1}	feet/sec.
miles/min.	1.6093	kms./min.
miles/min.	8.684×10^{-1}	knots/min.
miles/min.	6.0×10^{1}	miles/hr.
milliers	1.0×10^{3}	kilograms
millimicrons	1.0×10^{-9}	meters
milligrams	1.5432×10^{-2}	grains
milligrams	1.0×10^{-3}	grams
milligrams/liter	1.0	parts/million
millihenries	1.0×10^{-3}	henries
milliliters	1.0×10^{-3}	liters

To Convert	Multiply By	To Obtain
millimeters	1.0×10^{-1}	centimeters
millimeters	3.281×10^{-3}	feet
millimeters	3.937×10^{-2}	inches
millimeters	1.0×10^{-6}	kilometers
millimeters	1.0×10^{-3}	meters
millimeters	6.214×10^{-7}	miles
millimeters	3.937×10^{1}	mils
millimeters	1.094×10^{-3}	yards
million gals./day	1.54723	cu. ft./sec.
mils	2.54×10^{-3}	centimeters
mils	8.333×10^{-5}	feet
mils	1.0×10^{-3}	inches
mils	2.54×10^{-8}	kilometers
mils	2.778×10^{-5}	yards
miner's inches	1.5	cu. ft./min.
minims (british)	5.9192×10^{-2}	cubic cm.
minims (u.s. fluid)	6.1612×10^{-2}	cubic cm.
minutes (angles)	1.667×10^{-2}	degrees
minutes (angles)	1.852×10^{-4}	quadrants
minutes (angles)	2.909×10^{-4}	radians
minutes (angles)	6.0×10^{1}	seconds
minutes (time)	9.9206×10^{-5}	weeks
minutes (time)	6.944×10^{-4}	days
minutes (time)	1.667×10^{-2}	hours
minutes (time)	6.0×10^{1}	seconds
myriagrams	1.0×10^{1}	kilograms
myriameters	1.0×10^{1}	kilometers
myriawatts	1.0×10^{1}	kilowatts
	N	
nails	2.25	inches
newtons	1.0×10^{5}	dynes
	O	
ohm (international)	1.0005	ohm (absolute)
ohms	1.0×10^{-6}	megohms
ohms	1.0×10^{6}	microhms
ounces	8.0	drams
ounces	4.375×10^{2}	grains
ounces	2.8349×10^{1}	grams
ounces	6.25×10^{-2}	pounds
ounces	9.115×10^{-1}	ounces (troy)
ounces	2.790×10^{-5}	tons (long)
ounces	3.125×10^{-5}	tons (short)
ounces (fluid)	1.805	cu. inches
ounces (fluid)	2.957×10^{-2}	liters
ounces (troy)	4.80×10^{2}	grains
ounces (troy)	3.1103×10^{1}	grams
ounces (troy)	1.097	ounces (avdp.)
ounces (troy)	2.0×10^{1}	pennyweights (troy)
ounces (troy)	8.333×10^{-2}	pounds (troy)

To Convert	Multiply By	To Obtain
ounce/sq. in.	4.309×10^{3}	dynes/sq. cm.
ounce/sq. in.	6.25×10^{-2}	pounds/sq. in.
	P	
pace	3.0×10^{1}	inches
parsec	1.9×10^{13}	miles
parsec	3.084×10^{13}	kilometers
parts/million	5.84×10^{-2}	grains/u.s. gal.
parts/million	7.016×10^{-2}	grains/imp. gal.
parts/million	8.345	pounds/million gal.
pecks (british)	5.546×10^{2}	cubic inches
pecks (british)	9.0919	liters
pecks (u.s.)	2.5×10^{-1}	bushels
pecks (u.s.)	5.376×10^{2}	cubic inches
pecks (u.s.)	8.8096	liters
pecks (u.s.)	8	quarts (dry)
pennyweights (troy)	2.4×10^{1}	grains
pennyweights (troy)	5.0×10^{-2}	ounces (troy)
pennyweights (troy)	1.555	grams
pennyweights (troy)	4.1667×10^{-3}	pounds (troy)
pints (dry)	3.36×10^{1}	cubic inches
pints (dry)	1.5625×10^{-2}	bushels
pints (dry)	5.0×10^{-1}	quarts
pints (dry)	5.5059×10^{-1}	liters
pints (liquid)	4.732×10^{2}	cubic cms.
pints (liquid)	1.671×10^{-2}	cubic ft.
pints (liquid)	2.887×10^{1}	cubic inches
pints (liquid)	4.732×10^{-4}	cubic meters
pints (liquid)	6.189×10^{-4}	cubic yards
pints (liquid)	1.25×10^{-1}	gallons
pints (liquid)	4.732×10^{-1}	liters
pints (liquid)	5.0×10^{-1}	quarts (liquid)
planck's quantum	6.624×10^{-27}	erg-seconds
poise	1.0	gram/cm.-sec.
pounds (avdp.)	1.4583×10^{1}	ounces (troy)
poundals	1.3826×10^{4}	dynes
poundals	1.41×10^{1}	grams
poundals	1.383×10^{-3}	joules/cm.
poundals	1.383×10^{-1}	joules/meter (newtons)
poundals	1.41×10^{-2}	kilograms
poundals	3.108×10^{-2}	pounds
pounds	2.56×10^{2}	drams
pounds	4.448×10^{5}	dynes
pounds	7.0×10^{3}	grains
pounds	4.5359×10^{2}	grams
pounds	4.448×10^{-2}	joules/cm.
pounds	4.448	joules/meter (newtons)
pounds	4.536×10^{-1}	kilograms
pounds	1.6×10^{1}	ounces
pounds	1.458×10^{1}	ounces (troy)
pounds	3.217×10^{1}	poundals

To Convert	Multiply By	To Obtain
pounds	1.21528	pounds (troy)
pounds	5.0×10^{-4}	tons (short)
pounds (troy)	5.760×10^{3}	grains
pounds (troy)	3.7324×10^{2}	grams
pounds (troy)	1.3166×10^{1}	ounces (avdp.)
pounds (troy)	1.2×10^{1}	ounces (troy)
pounds (troy)	2.4×10^{2}	pennyweights (troy)
pounds (troy)	8.2286×10^{-1}	pounds (avdp.)
pounds (troy)	3.6735×10^{-4}	tons (long)
pounds (troy)	3.7324×10^{-4}	tons (metric)
pounds (troy)	4.1143×10^{-4}	tons (short)
pounds of water	1.602×10^{-2}	cu. ft.
pounds of water	2.768×10^{1}	cu. inches
pounds of water	1.198×10^{-1}	gallons
pounds of water/min.	2.670×10^{-4}	cu. ft./sec.
pound-feet	1.356×10^{7}	cm.-dynes
pound-feet	1.3825×10^{4}	cm.-grams
pound-feet	1.383×10^{-1}	meter-kgs.
pounds/cu. ft.	1.602×10^{-2}	grams/cu. cm.
pounds/cu. ft.	1.602×10^{1}	kgs./cu. meter
pounds/cu. ft.	5.787×10^{-4}	pounds/cu. inches
pounds/cu. ft.	5.456×10^{-9}	pounds/mil-foot
pounds/cu. in.	2.768×10^{1}	grams/cu. cm.
pounds/cu. in.	2.768×10^{4}	kgs./cu. meter
pounds/cu. in.	1.728×10^{3}	pounds/cu. ft.
pounds/cu. in.	9.425×10^{-6}	pounds/mil-foot
pounds/ft.	1.488	kgs./meter
pounds/in.	1.786×10^{2}	grams/cm.
pounds/mil-foot	2.306×10^{6}	grams/cu. cm.
pounds/sq. ft.	4.725×10^{-4}	atmospheres
pounds/sq. ft.	1.602×10^{-2}	feet of water
pounds/sq. ft.	1.414×10^{-2}	inches of mercury
pounds/sq. ft.	4.882	kgs./sq. meter
pounds/sq. ft.	6.944×10^{-3}	pounds/sq. inch
pounds/sq. in.	6.804×10^{-2}	atmospheres
pounds/sq. in.	2.307	feet of water
pounds/sq. in.	2.036	inches of mercury
pounds/sq. in.	7.031×10^{2}	kgs./sq. meter
pounds/sq. in.	1.44×10^{2}	pounds/sq. ft.
pounds/sq. in.	7.2×10^{-2}	short tons/sq. ft.
pounds/sq. in.	7.03×10^{-2}	kgs./sq. cm.
	Q	
quadrants (angle)	9.0×10^{1}	degrees
quadrants (angle)	5.4×10^{3}	minutes
quadrants (angle)	1.571	radians
quadrants (angle)	3.24×10^{5}	seconds
quarts (dry)	6.72×10^{1}	cu. inches
quarts (liquid)	9.464×10^{2}	cu. cms.
quarts (liquid)	3.342×10^{-2}	cu. ft.
quarts (liquid)	5.775×10^{1}	cu. inches

To Convert	Multiply By	To Obtain
quarts (liquid)	9.464×10^{-4}	cu. meters
quarts (liquid)	1.238×10^{-3}	cu. yards
quarts (liquid)	2.5×10^{-1}	gallons
quarts (liquid)	9.463×10^{-1}	liters
	R	
radians	5.7296×10^{1}	degrees
radians	3.438×10^{3}	minutes
radians	6.366×10^{-1}	quadrants
radians	2.063×10^{5}	seconds
radians/sec.	5.7296×10^{1}	degrees/sec.
radians/sec.	9.549	revolutions/min.
radians/sec.	1.592×10^{-1}	revolution/sec.
radians/sec./sec.	5.7296×10^{2}	revs./min./min.
radians/sec./sec.	9.549	revs./min./sec.
radians/sec./sec.	1.592×10^{-1}	revs./sec./sec.
reams	5.0×10^{2}	sheets
revolutions	3.60×10^{2}	degrees
revolutions	4.0	quadrants
revolutions	6.283	radians
revolutions/min.	6.0	degrees/sec.
revolutions/min.	1.047×10^{-1}	radians/sec.
revolutions/min.	1.667×10^{-2}	revs./sec.
revs./min./min.	1.745×10^{-3}	radians/sec./sec.
revs./min./min.	1.667×10^{-2}	revs./min./sec.
revs./min./min.	2.778×10^{-4}	revs./sec./sec.
revolutions/sec.	3.6×10^{2}	degrees/sec.
revolutions/sec.	6.283	radians/sec.
revolutions/sec.	6.0×10^{1}	revs./min.
revs./sec./sec.	6.283	radians/sec./sec.
revs./sec./sec.	3.6×10^{3}	revs./min./min.
revs./sec./sec.	6.0×10^{1}	revs./min./sec.
rods	2.5×10^{-1}	chains (gunters)
rods	5.029	meters
rods (surveyors' meas.)	5.5	yards
rods	1.65×10^{1}	feet
rods	1.98×10^{2}	inches
rods	3.125×10^{-3}	miles
rope	2.0×10^{1}	feet
	S	
scruples	2.0×10^{1}	grains
seconds (angle)	2.778×10^{-4}	degrees
seconds (angle)	1.667×10^{-2}	minutes
seconds (angle)	3.087×10^{-6}	quadrants
seconds (angle)	4.848×10^{-6}	radians
slugs	1.459×10^{1}	kilograms
slugs	3.217×10^{1}	pounds
sphere (solid angle)	1.257×10^{1}	steradians
square centimeters	1.973×10^{5}	circular mils
square centimeters	1.076×10^{-3}	sq. feet

To Convert	Multiply By	To Obtain
square centimeters	1.550×10^{-1}	sq. inches
square centimeters	1.0×10^{-4}	sq. meters
square centimeters	3.861×10^{-11}	sq. miles
square centimeters	1.0×10^{2}	sq. millimeters
square centimeters	1.196×10^{-4}	sq. yards
square degrees	3.0462×10^{-4}	steradians
square feet	2.296×10^{-5}	acres
square feet	1.833×10^{8}	circular mils
square feet	9.29×10^{2}	sq. cms.
square feet	1.44×10^{2}	sq. inches
square feet	9.29×10^{-2}	sq. meters
square feet	3.587×10^{-8}	sq. miles
square feet	9.29×10^{4}	sq. millimeters
square feet	1.111×10^{-1}	sq. yards
square inches	1.273×10^{6}	circular mils
square inches	6.452	sq. cms.
square inches	6.944×10^{-3}	sq. ft.
square inches	6.452×10^{2}	sq. millimeters
square inches	1.0×10^{6}	sq. mils
square inches	7.716×10^{-4}	sq. yards
square kilometers	2.471×10^{2}	acres
square kilometers	1.0×10^{10}	sq. cms.
square kilometers	1.076×10^{7}	sq. ft.
square kilometers	1.550×10^{9}	sq. inches
square kilometers	1.0×10^{6}	sq. meters
square kilometers	3.861×10^{-1}	sq. miles
square kilometers	1.196×10^{6}	sq. yards
square meters	2.471×10^{-4}	acres
square meters	1.0×10^{4}	sq. cms.
square meters	1.076×10^{1}	sq. ft.
square meters	1.55×10^{3}	sq. inches
square meters	3.861×10^{-7}	sq. miles
square meters	1.0×10^{6}	sq. millimeters
square meters	1.196	sq. yards
square miles	6.40×10^{2}	acres
square miles	2.788×10^{7}	sq. ft.
square miles	2.590	sq. kms.
square miles	2.590×10^{6}	sq. meters
square miles	3.098×10^{6}	sq. yards
square millimeters	1.973×10^{3}	circular mills
square millimeters	1.0×10^{-2}	sq. cms.
square millimeters	1.076×10^{-5}	sq. ft.
square millimeters	1.55×10^{-3}	sq. inches
square mils	1.273	circular mils
square mils	6.452×10^{-6}	sq. cms.
square mils	1.0×10^{-6}	sq. inches
square yards	2.066×10^{-4}	acres
square yards	8.361×10^{3}	sq. cms.
square yards	9.0	sq. ft.
square yards	1.296×10^{3}	sq. inches
square yards	8.361×10^{-1}	sq. meters

To Convert	Multiply By	To Obtain
square yards	3.228×10^{-7}	sq. miles
square yards	8.361×10^{5}	sq. millimeters
steradians	7.958×10^{-2}	spheres
steradians	1.592×10^{-1}	hemispheres
steradians	6.366×10^{-1}	spherical right angles
steradians	3.283×10^{3}	square degrees
steres	9.99973×10^{2}	liters
	T	
temperature (°C.) +273	1.0	absolute temperature (°K.)
temperature (°C.) +17.78	1.8	temperature (°F.)
temperature (°F.) +460	1.0	absolute temperature (°R.)
temperature (°F.) −32	5/9	temperature (°C.)
tons (long)	1.016×10^{3}	kilograms
tons (long)	2.24×10^{3}	pounds
tons (long)	1.12	tons (short)
tons (metric)	1.0×10^{3}	kilograms
tons (metric)	2.205×10^{3}	pounds
tons (short)	9.0718×10^{2}	kilograms
tons (short)	3.2×10^{4}	ounces
tons (short)	2.9166×10^{4}	ounces (troy)
tons (short)	2.0×10^{3}	pounds
tons (short)	2.43×10^{3}	pounds (troy)
tons (short)	8.9287×10^{-1}	tons (long)
tons (short)	9.078×10^{-1}	tons (metric)
tons (short)/sq. ft.	9.765×10^{3}	kgs./sq. meter
tons (short)/sq. ft.	1.389×10^{1}	pounds/sq. in.
tons (short)/sq. in.	1.406×10^{6}	kgs./sq. meter
tons (short)/sq. in.	2.0×10^{3}	pounds/sq. in.
tons of water/24 hrs.	8.333×10^{1}	pounds of water/hr.
tons of water/24 hrs.	1.6643×10^{-1}	gallons/min.
tons of water/24 hrs.	1.3349	cu. ft./hr.
	V	
volt/inch	3.937×10^{7}	abvolts/cm.
volt/inch	3.937×10^{-1}	volt/cm.
volt (absolute)	3.336×10^{-3}	statvolts
volts	1.0×10^{8}	abvolts
	W	
watts	3.4129	btu/hr.
watts	5.688×10^{-2}	btu/min.
watts	1.0×10^{7}	ergs/sec.
watts	4.427×10^{1}	ft.-lbs./min.
watts	7.378×10^{-1}	ft.-lbs./sec.
watts	1.341×10^{-3}	horsepower
watts	1.36×10^{-3}	horsepower (metric)
watts	1.433×10^{-2}	kg.-calories/min.
watts	1.0×10^{-3}	kilowatts
watts (abs.)	1.0	joules/sec.
watt-hours	3.413	btu

To Convert	Multiply By	To Obtain
watt-hours	3.6×10^{10}	ergs
watt-hours	2.656×10^{3}	foot-lbs.
watt-hours	8.605×10^{2}	gram-calories
watt-hours	1.341×10^{-3}	horsepower-hours
watt-hours	8.605×10^{-1}	kilogram-calories
watt-hours	3.672×10^{2}	kilogram-meters
watt-hours	1.0×10^{-3}	kilowatt-hours
watt (international)	1.000165	watt (absolute)
webers	1.0×10^{8}	maxwells
webers	1.0×10^{5}	kilolines
webers/sq. in.	1.55×10^{7}	gausses
webers/sq. in.	1.0×10^{8}	lines/sq. in.
webers/sq. in.	1.55×10^{-1}	webers/sq. cm.
webers/sq. in.	1.55×10^{3}	webers/sq. meter
webers/sq. meter	1.0×10^{4}	gausses
webers/sq. meter	6.452×10^{4}	lines/sq. in.
webers/sq. meter	1.0×10^{-4}	webers/sq. cm.
webers/sq. meter	6.452×10^{-4}	webers/sq. in.
weeks	1.68×10^{2}	hours
weeks	1.008×10^{4}	minutes
weeks	6.048×10^{5}	seconds
	Y	
yards	9.144×10^{1}	centimeters
yards	9.144×10^{-4}	kilometers
yards	9.144×10^{-1}	meters
yards	4.934×10^{-4}	miles (nautical)
yards	5.682×10^{-4}	miles (statute)
yards	9.144×10^{2}	millimeters
years	3.65256×10^{2}	days (mean solar)
years	8.7661×10^{3}	hours (mean solar)

USEFUL PHYSICAL CONSTANTS

Gas Constants (R)

R = 0.0821	(atm.) (liter)/(g.-mole) (°K)
R = 1.987	g.-cal./(g.-mole) (°K)
R = 1.987	B.t.u./(lb.-mole) (°R)
R = 1.987	c.h.u./(lb.-mole) (°K)
R = 8.314	joules/(gm-mole) (°K)
R = 1,546	(ft.) (lb.force)/(lb.-mole) (°R)
R = 10.73	(lb.-force/sq.in.) (cu.ft.)/(lb.-mole) (°R)
R = 18510	(lb.-force/sq.in.) (cu.in.)/(lb.-mole) (°R)
R = 0.7302	(atm.) (cu.ft.)/(lb.-mole) (°R)
R = 8.48×10^{5}	(Kg./m^2) (cu.cm.)/(lb.-mole) (°K)

Acceleration of Gravity (Standard)

$g = 32.17$ ft./sec.^{2} = 980.6 cm./sec.^{2}

Velocity of Sound in Dry Air @ 0°C and 1 atm.

33,136 cm./sec. = 1,089 ft./sec.

Heat of Fusion of Water
79.7 cal./gm = 144 Btu/lb.
Heat of Vaporization of Water @ 1.0 atm.
540 cal./gm = 970 Btu/lb.
Specific Heat of Air
Cp = 0.238 cal./(gm)(°C)
Density of Dry Air @ 0°C and 760 mm.
0.001293 gm/cu.cm.

REFERENCES

1. Garabedian, P. R. *Partial Differential Equations*. New York: John Wiley & Sons, 1964.
2. Guillemin, E. A. *The Mathematics of Circuit Analysis*. New York: John Wiley & Sons, 1958.
3. Hsu, H. P. *Outline of Fourier Analysis*. New York: Associated Educational Services Corp., UNITECH DIVISION, 1967.
4. Kirchmayer, L. K. *Economic Operation of Power Systems*. New York: John Wiley & Sons, 1958.
5. Kuo, B. C. *Automatic Control Systems,* 4th ed. Englewood Cliffs, N.J.: Prentice Hall, 1982.
6. Oberhettinger, F., and L. Badii. *Tables of Laplace Transforms*. New York, Heidelberg, Berlin: Springer-Verlag.
7. Papoulis, A. *Circuit and Systems, A Modern Approach*. New York: Holt, Rinehart and Winston, 1980.
8. Taxel, Irving. *Conversion Factors*. Newark, N.J.: NJIT Alumni Association, Copyright 1964.

Index